SpringerBriefs in Latin American Studies

More information about this series at http://www.springer.com/series/14332

Martha G. Bell

Pottery, Livelihoods, and Landscapes

A Case Study from the Peruvian Andes

CLAG
clagscholar.org

Springer

Martha G. Bell
Department of Humanities,
Geography Section
Pontifical Catholic University of Peru
Lima
Peru

ISSN 2366-763X ISSN 2366-7648 (electronic)
SpringerBriefs in Latin American Studies
ISBN 978-3-319-52330-9 ISBN 978-3-319-52331-6 (eBook)
DOI 10.1007/978-3-319-52331-6

Library of Congress Control Number: 2017932399

Printed on acid-free paper

This Springer imprint is published by Springer Nature
The registered company is Springer International Publishing AG
The registered company address is: Gewerbestrasse 11, 6330 Cham, Switzerland

Micaela Jiménez, Orfilia Mondragón, Mauro Mondragón, Irma Abad, Daniel Mondragón, Emma Mondragón, Hortensia Criollo y todos los comuneros de San Bartolomé de los Olleros.

Acknowledgements

This study of pottery exchange grew out of a much larger project, directed by Gabriel Ramón (now professor at the Pontifical Catholic University of Peru), on pottery production in the Northern Peruvian Andes. I am deeply indebted to Gabriel for allowing me to participate in his work and for supporting my interest in pottery trade and distribution. I would also like to thank friend and fellow project member José Luis Pino for introducing me to both the Andean highlands and pottery communities on one memorable weekend in 2002.

I was lucky to pursue this project while a graduate student in the Department of Geography of the University of Wisconsin-Madison. I am grateful to my teachers there, especially Karl Zimmerer, Matt Turner, and Frank Salomon, for their insights and encouragement. Members of Geography's Human Environment Research Discussion group (HERD), as well as fellow Andeanist graduate students, including Tony Chapa, Claudia Ochoa, and Dustin Welch, provided meaningful contributions. At UW-Madison I also thank Armando Muyulema, Ariana Hauck, and Rob Roth for Kichwa instruction, discussion of indigenous livelihoods, and cartography training, respectively. This project was funded by a Tinker-Nave Short-Term Field Research Grant from Latin American, Caribbean, and Iberian Studies at UW-Madison.

In Lima, my research was supported by Hildegardo Córdova of the Center for Applied Geographic Research at the Pontifical Catholic University of Peru.

In Piura, I am grateful to the staff of the CIPCA library.

For their assistance (and company) during the writing period, I would like to thank Nora Ramón, Andrés Ramón, Gabriel Ramón, and Ysabel Vilcabana. Andrew Sluyter and one anonymous reviewer provided helpful comments and corrections. Stephanie Orrego updated my original maps with recently digitized data.

Finalmente, muy agradecida a Micaela Jiménez, Orfilia Mondragón, Mauro Mondragón, Irma Abad, Daniel Mondragón, Emma Mondragón, Hortensia Criollo y todos los comuneros de San Bartolomé de los Olleros.

Contents

1 Approaches to the Study of Pottery Exchange and Rural Livelihoods in San Bartolomé de los Olleros: Concepts, Background, and Methods 1
1.1 Introduction 1
1.2 The Andean Studies Perspective 2
1.3 The View from Development Studies and the "New Rurality" 5
1.4 Andean Pottery Production and Trade 7
1.5 Pottery Production in Piura 8
1.6 Pottery Production in San Bartolomé de los Olleros 11
1.7 Research Methods 18
1.7.1 Data Collection 18
1.7.2 Data Analysis 20
References 22

2 Networks of Pottery Exchange 25
2.1 Methods of Pottery Distribution and the Case of San Bartolomé de los Olleros 25
2.2 Trading Trips 27
2.3 Community-Based Exchange 35
2.4 Visiting Customers: Ad Hoc Trade from the Potter's House and Bulk Purchase 36
2.5 Annual Fairs 39
2.6 Itinerant Potters and Peonage 40
2.7 Urban Markets 41
2.8 Conclusions 42
References 42

3 Negotiating the Pottery Exchange Landscape 43
3.1 Kinds of Places, Kinds of Food: The Geography of Agriculture in Piura 43
3.1.1 Trading Pots for Food. 43
3.1.2 Ecology and Agriculture. 46
3.1.3 Cuisine and Complementarity. 50
3.2 Exchange Rates and Methods of Measurement. 52
3.2.1 Exchange Rates 52
3.2.2 Measurement Systems. 56
3.3 Potters' Life Cycles and Trade Practices. 58
3.4 Conclusions. 59
References. 59

4 Pottery Exchange and Livelihoods: An Assessment 61
References. 67

Appendix A: Profiles of the Potters and Traders Interviewed. 69

Appendix B: Interview Questions and Topics. 71

Appendix C: Registry of Trade Routes, Destinations, and Customer Points of Origin 75

Appendix D: Registry of Products, Places, and Exchange Rates 81

Index 87

Introduction: Pottery Exchange and the Spatiality of Rural Livelihoods

In 2004 a project co-sponsored by the United Nations Development Program[1] whose goal was the improvement of handicraft production, specifically decorative ceramics, was implemented in San Bartolomé de los Olleros, a rural highland community located in the extreme northern sierra of Peru (Department of Piura). This project drew on Olleros' long tradition of pottery production, and using what is now common rhetoric, sought to support "an alternative activity for sustainable economic development" (UNDP et al. 2004, my translation). A workshop was constructed, a kiln was built, and local women were trained to make ceramic figurines using molds with shapes based on the "anthropomorphic designs of petroglyphs originating from their ancestral culture" (*ibid*). These decorative figurines were oriented towards the national tourist economy, and the enterprise was supposed to help the women develop "efficient use of their natural resources" (*ibid*). (Fig. 1). Yet by 2005, the molds shaped in the forms of ashtrays, vases, and other trinkets were lying abandoned in piles on the workshop floor, the workshop itself had been converted into a meeting center for the local Mother's Club (*Club de Madres*), and the kiln was obscured by the overgrowth of tall grasses and weeds.

Meanwhile, normal pottery production in Olleros, which is the elaboration of utilitarian clay vessels for cooking and storage of foods and beverages, continued as usual. While perhaps it was quite predictable that such a project would not be successful, it is interesting to consider some of the reasons why it did fail. In fact, such reasons are numerous. The project taught local women a ceramic production technique completely distinct from the tradition that has existed in Olleros for

[1]"Las Ceramistas de Olleros" Project was jointly sponsored by UNDP (Perú), UNOPS, SGP (Programa de Pequeñas Donaciones), FMAM, and IGCH (Instituto de Gestión de Cuencas Hidrográficas). The information about this project presented here was drawn from conversations with local participants (potters) as well as promotional materials advertising the project. Analysis of the SGP website, http://www.sgp.undp.org, in 2016 shows 24 projects worldwide dealing with ceramics since 1996, but does not list the Olleros project. Similarly, a current employee at UNOPS could find no record or memory of the project.

Fig. 1 Promotional poster for the "Ceramistas de Olleros" project

generations. The technique called for a variety of clay not available locally and thus not easily acquired. And perhaps most importantly, the finished products, which were figurines in the same style as those made in the massive production centers of tourist goods on the coast (e.g., Chulucanas), never really had a chance to compete in an already glutted market.

Much more interesting than merely recounting such faults is to consider the broader assumptions that this type of project made about the role of pottery in the livelihoods of Olleros' residents.[2] These assumptions relate to both the function of markets in Olleros livelihoods as well as to the spatiality of pottery exchange as a livelihood activity. In fact, the way this project conceived of pottery: as an activity used to produce an object for market sale and thus to derive cash income, was not at all the way that Olleros residents actually use pots. In Olleros, as in many parts of the Andes, pots are mainly bartered for food crops within a nonmonetary peasant economy distinct from the "conventional" market. Therefore, this book explores the basic questions: How is pottery used as a livelihood activity among the residents of Olleros? Why is barter of pots practiced? And how might pottery production and exchange practices be changing? To discuss these themes, the following chapters will analyze links between handicraft production, rural trade networks, and agricultural practices in the Andean context.

---

[2]For other examples of ceramic development projects see Mohr (1992: 64) on the Raqch'i potters in Cuzco, Peru, or Echeandia (1982) on projects in Ancash, Peru.

Recent literature in the broadly defined field of political ecology has critically discussed community-based conservation and development projects, including those related to payments for ecological services and to climate change mitigation (there are many examples, but see Agrawal and Gibson 1999; Bebbington 2000, 2001; Evans, Murphy and de Jong 2014; Himley 2009; Perreault 2009). Drawing on earlier insights from cultural ecologists such as Netting, who argued that foreign development projects (at least in Africa) tended to ignore "the existence of working indigenous solutions to the problems of farming an alien environment" (Netting 1993: 22); much of the focus of this work has been to evaluate how well these projects fit (or do not fit) with the ecological, social, and political dynamics of the communities in which they are applied. Significantly, many of them reach their conclusions through the detailed analysis of local livelihoods. These types of studies are a good starting point for beginning to break down the assumptions made by the ceramic development project described above, and for understanding the actual ways in which pottery trade serves as a livelihood activity.

The first assumption has to do with misunderstandings about the ways that pots are marketed and the role of markets in rural highland Andean communities like Olleros. In their discussion of cattle markets in the Sahel, Turner and Williams (2002: 683) conclude that "rural markets may function quite differently than assumed by development practitioners." In fact, these "*real markets*" are often dominated by institutional, infrastructural, cultural, and even seasonal and ecological conditions that create patterns that may bear little resemblance to rational economic predictions. Such insights are certainly relevant to informal systems of pottery exchange. Perhaps this is especially true when nonmonetary barter is the most prevalent mode of exchange, as it often is for pottery. This is primarily because such barter may not occur in defined marketplaces and customary practices for controlling rates of exchange (prices) may not be immediately apparent. So if the objective is to gain awareness of how pottery exchange actually works as a livelihood activity, the first step must be to examine closely the various circumstances in which pots are traded. This entails looking at both the material goods being exchanged as well as the range of forces—political-economic, historic, socio-cultural and/or environmental—that create the institutional context in which exchange occurs. From this foundation, we have the basis not only for understanding pottery exchange, but also for approaching broader trade and market dynamics in the rural Andes.

A second assumption relates to the spatiality of pottery production as a livelihood activity, as expressed by the actual geographic patterns of production and exchange. Spatiality here is defined by Zimmerer (2006a: 9, referencing Gregory 2000: 717) as "configurations in which 'physical extent' is 'fused with social intent.'" Spatiality has been deemed central to livelihood analyses "that seek to understand not only the dynamic, extra-local reality of most livelihood practices, but also how the production and reproduction of livelihoods are interlinked with the processes of producing and reproducing space" (King 2011: 298). While Olleros pots are used within the immediate locale, by the potters themselves as well as by their neighbors, the pots are mainly produced for trade. They are exchanged with consumers in the surrounding region, the majority of whom are rural people living in farming or herding-based *campesina* communities.[3] This trade represents a significant component of pottery traders' income (in terms of cash and bartered produce), yet exchanges almost always occur informally between individual households across multiple communities. In many cases, trade depends on long-term interactions and relationships between individuals, farmers, and communities. Thus pursuing pottery trade means drawing upon, and continuously reproducing, social, and spatial relationships, full of cultural and historical, as well as personal or individual, meaning. To fully comprehend these patterns, both the local and the extra-local, we must take a scalar approach which privileges the examination of the multi-community network over a focus on the territory of a single community. By analyzing flows of goods, but also flows of people and information, we are quickly reminded that all communities are connected to surrounding places, both near and far, and thus cannot be considered as independent entities.

One way to simultaneously address both the issues of real markets and the spatiality of livelihood practices is through the analysis of the trade networks through which people acquire goods of everyday importance. McSweeney (2004: 639), in a geographic study of the trade of dugout canoes in the Mosquitia region of Honduras, has presented a useful model for such an analysis. She demonstrates that "attention to rural trade networks… can offer researchers and practitioners a parsimonious means to simultaneous engage rural peoples' everyday economic imperatives while giving due weight to the multi-scaled and multi-sited social and political processes through which rural livelihoods are constituted." McSweeney's methodology of following a highly visible material object as it is traded between sites as a means of identifying and visualizing the multi-sited networks through which livelihoods are created is highly applicable to this study of pottery exchange. Pots, like canoes, are easily identifiable by place of origin, and even frequently by potter, and thus it is relatively straightforward to visualize networks of exchange, even without the testimonies of producers, traders, or consumers. This reality, of course, has proven useful for archaeologists (see more discussion in Chap. 4). Other recent studies, for example Turner (2007) work on textile traders in Vietnam, Zimmerer (2006b) study of potato seed networks in Peru, and Coomes (2010) research on crop plant material exchange also highlight the importance of tracing flows of goods as a means of understanding the spatiality of livelihoods, and significantly, together with McSweeney's analysis, these studies remind us that this

[3]There is a fairly complex history of peasant communities in Peru. Beginning in the colonial era *Comunidades* were recognized. Later in 1919, the Peruvian constitution officially recognized Indigenous communities (*Comunidades Indígenas*), and most recently, after the Agrarian Reform (1969), *Comunidades Campesinas* were legally formed from ex-Haciendas. There are 6000 *comunidades campesinas* in Peru (see http://www.allpa.org.pe/).

spatiality of livelihoods shifts depending on product. Pots, cloth, seeds, plant clippings, and canoes each flow across their own unique networks. Perreault (2009) rightly emphasizes that such spatiality can incorporate both the rural and urban, and that rather than being restricted to a local scale, livelihood networks often expand even to the transnational.

This book takes a potters'-eye view of the dynamic natural, social, and economic landscape that potters must negotiate to acquire goods of everyday importance. This does not mean to imply that the research presented here is the result of deep ethnographic study. Rather, that the potters themselves are the center of the analysis of pottery distribution. Of primary importance are: (1) the decision making processes surrounding exchange activities, (2) how exchange choices derive from and result in distinct spatial patterns (in other words, how choices produce and reproduce the spatiality of livelihoods), and (3) how the marketing of pots impacts potter livelihoods. This potter-centric approach contrasts with ceramic analyses that are more interested in the distribution of the material object than in the livelihood of the artisan. But in fact, the two components are intertwined. In order to understand how pottery is used as livelihood activity, this book considers several more specific questions: Why does distribution occur in the way it does? How and why do pots move from one place to another? And, finally, why is pottery production useful for potters?

Such a substantivist approach—one that considers the physical, cultural, and social features of economic systems as interconnected and continuous analytic categories (Halperin 1977: 6), embedded within each other and within exchange practices (Hess 2004: 169), leads to more realistic notions of the actual use of pottery as a livelihood activity. Thus more broadly, this book seeks to expand the discussion of small-scale trade as an important component of rural Andean livelihoods, especially through attention to the spatiality of this practice. Since pots are produced to be traded, and moreover, since they are mainly produced to be bartered for food crops, the resulting trade patterns stem directly from regional agropastoral organization. When planning trading activities potters must determine not only what products they seek, but also where those products are grown and when they are seasonally available. In addition, they must consider the exchange situations they will encounter in each of these locations, including both the accepted equivalencies of farm products for pots, as well as their interactions with trading partners over time. This book, therefore, seeks to combine the geography of agriculture with the geography of exchange, taking into account the relevant environmental, social, cultural, and economic components. It aims to ethnographically demonstrate how individual pottery traders negotiate these intersecting forces, creating networks of production and exchange as they earn their livings. In doing so, it shows why pottery exchange (and barter more generally) continues to be useful, and why it persists in spite of ever-increasing market integration. When examined from this perspective, pottery production as an income activity begins to look very different from how it was conceived by the UNDP ceramic development plan.

References

Agrawal A, Gibson CC (1999) Enchantment and disenchantment: the role of community in natural resource conservation. World Dev 27(4):629–649

Bebbington A (2000) Reencountering development: livelihood transitions and place transformations in the Andes. Ann Assoc Am Geogr 90(3):495–520

Bebbington A (2001) Globalized Andes? Livelihoods, landscapes and development. Ecumene 8(4):414–436

Coomes OT (2010) Of stakes, stems, and cuttings: the importance of local seed systems in traditional Amazonian societies. Prof Geogr 62(3):323–334

Echeandia JM (1982) Alfareria tradicional en Taricá (Ancash). Lima: Historia Andina. Seminario de Historia Rural Andina

Evans K, Murphy L, de Jong W (2014) Global versus local narratives of REDD: a case study from Peru's Amazon. Environ Sci Policy 35:98–108

Gregory D (2000) Spatiality. In: Johnston RJ, Gregory D, Pratt G et al (eds) The dictionary of human geography, 4th edn. Blackwell, Oxford, pp 715–717

Halperin R (1977) Introduction: the substantive economy in peasant societies. In: Halperin R, Dow J (eds) Peasant livelihood. St. Martin's Press, New York, pp 1–16

Hess M (2004) 'Spatial' relationships? Towards a reconceptualization of embeddedness. Prog Hum Geogr 28(2):165–186

Himley M (2009) Nature conservation, rural livelihoods, and territorial control in Andean Ecuador. Geoforum 40(5):832–842

King B (2011) Spatialising livelihoods: resource access and livelihood spaces in South Africa. Trans Inst Brit Geogr 36:297–313

McSweeney K (2004) The dugout canoe trade in Central America's Mosquitia: Approaching rural livelihoods through systems of exchange. Ann Assoc Am Geogr 94(3):638–661

Mohr KL (1992) The organization of production and distribution of traditional pottery in South Highland Peru. In: Bey GJ III, Pool CA (eds) Ceramic production and distribution. Westview Press, Boulder, pp 49–92

Netting R McC (1993) Smallholders, householders: farm families and the ecology of intensive, sustainable agriculture. Stanford University Press, Palo Alto

Perreault T (2009) Environment and development. In: Castree N, Demeritt D, Liverman et al (eds) A companion to environmental geography. Wiley, Malden, pp 442–460

Turner S (2007) Trading old textiles: the selective diversification of highland livelihoods in Northern Vietnam. Hum Organ, 66(4):389–404

Turner MD, Williams TO (2002) Livestock market dynamics and local vulnerabilities in the Sahel. World Dev 30(4):683–705

UNDP, UNOPS, SGP, FMAM, IGCH (2004) "Las Ceramistas de Olleros" Promotional Poster

Zimmerer KS (2006a) Geographical perspectives on globalization and environmental issues: the inner-connections of conservation, agriculture, and livelihoods. In: Zimmerer KS (ed) Globalization and new geographies of conservation. University of Chicago Press, Chicago, pp 1–43

Zimmerer KS (2006b) Multilevel geographies of seed networks and seed use in relation to agrobiodiversity conservation in the Andean Countries. In: Zimmerer KS (ed) Globalization and new geographies of conservation. University of Chicago Press, Chicago, pp 141–165

Chapter 1
Approaches to the Study of Pottery Exchange and Rural Livelihoods in San Bartolomé de los Olleros: Concepts, Background, and Methods

Abstract This chapter presents the concepts, background information, and methods used to study the livelihoods and exchange practices of the potters of San Bartolomé de los Olleros. It draws on theory, concepts, and advances from the two broad fields of Andean Studies and Development Studies. In Andean Studies, it particularly considers the themes of ecological complementarity and exchange. From Development Studies, it engages with concepts of diversification of livelihoods and nonagricultural or off-farm rural livelihood activities. This study is also placed in the context of studies of pottery production in the Andes more generally. Specifically, it stems from a larger survey of pottery production in the Northern Peruvian Andes (see especially Ramón 2008; Ramón and Bell 2013), which provides important background information about regional practices. This chapter also includes a basic introduction to pottery production in Olleros, and explains the relationship between pottery and other livelihood activities. Finally, the chapter concludes with a presentation of the research and analytic methods used in this study.

Keywords Livelihoods · Diversification · Ecological complementarity · Pottery · Andes

1.1 Introduction

Potter livelihoods can be approached from a number of perspectives. This case study of pottery production and exchange in San Bartolomé de los Olleros (Piura, Peru) considers these activities through two lenses. The first is that of Andean Studies, a diverse field encompassing anthropology, geography, archaeology, and history, which has made significant contributions to understandings of Andean lifestyles in terms of ecological complementarity and the social, cultural, and economic components of production and exchange practices. The second is that of Development Studies, perhaps an even broader field, which seeks to understand, among other things, rural livelihoods, and their change in Latin America and the Global South more generally. Here, insights about livelihood diversification and nonagricultural or

M.G. Bell, *Pottery, Livelihoods, and Landscapes*, SpringerBriefs in Latin American Studies, DOI 10.1007/978-3-319-52331-6_1

off-farm rural activities are considered in the context of a globalizing economy and the discussion of the so-called "New Rurality."

The Olleros case must also be understood in comparison with other studies of pottery production and exchange in the Andean region generally, and in the Department of Piura specifically. To this end, this chapter includes detailed background information and description of pottery production in Olleros, especially in relation to the other livelihood activities pursued in the community. Finally, this study relies on a qualitative research methodology based on semi-structured interviews, participant observation, and mapping and spatial analysis of exchange patterns. The remainder of this chapter is devoted to explaining these conceptual, contextual, and methodological features.

1.2 The Andean Studies Perspective

In Andean Studies there is a long research tradition relating to spatial conceptions of livelihoods, particularly with respect to Murra's (2002 [1972]) theory of vertical ecological complementarity. Murra's basic approach, as succinctly summarized by Salomon (1985: 511), is that "for geographical reasons the Andean people must achieve the levels of consumption defined by their cultures as adequate through the articulation of complementary productive zones at varied altitudes and distances." According to the verticality concept, the Andean ideal would be to have direct access to lands in as many of these productive zones as possible. However, given that such access is difficult to achieve, over time Andean peoples have developed various exchange practices to acquire goods they do not produce for themselves. Thus exchange has often been understood as a mechanism for transferring or distributing products between disparate ecological zones, which are usually defined by a difference in elevation.

One classic example of such complementarity is the interrelationship between high-altitude pastoralists and middle elevation farmers who trade animal products for maize, potatoes, and other crops (Flores Ochoa 1977; Fonseca 1973; Göbel 1998; Tomoeda 1985). However, the products involved in such interzonal exchanges vary considerably, including everything from coca to wheat and from fruits to freshwater fish. Likewise, ecological complementarity is not exclusively driven by elevation-driven variation. The uneven geography of the Andean landscape leads to regional differences in numerous features (e.g., temperature, soil quality, water availability), which result in variation in yields and timing of harvests as well as in specialization of production. These traits have also been identified as drivers of exchange (Mayer 2002: 148), although as Mayer (1985) argues, the focus should be on "Production Zones," since ecological features are in reality mediated by social relations and institutions (e.g., irrigation systems, land tenure patterns).

Additionally, many strategies for moving goods between zones are employed by Andean people (for a classification see Salomon 1985). Exchange can take place in contexts ranging from annually organized caravans to microscale barter between individuals, or from itinerant merchants peddling goods for profit to centralized marketplaces with many sellers hawking their wares (Fonseca 1973; Mayer 2002). Not unexpectedly, in each of these contexts traders encounter different social relationships, which can manifest themselves in terms of bargaining power, form of payment (cash or kind), and the types of products exchanged (industrial or *campesino* production) (see also Weismantel 2001: 144–6).

A key feature to consider is that in any of these circumstances the actual exchange may occur either by sale/purchase with cash or by barter of product for product. To explain this occurrence in the Andean context, Mayer (2002: 158) has postulated the existence of two separate spheres of exchange: the market (monetary) and the nonmarket (nonmonetary reciprocal exchange or barter). He argues that rural Andean people often deliberately choose to avoid exploitation in the market—physical marketplaces and the market economy more generally—by participating in nonmonetary barter at the individual level. He further argues that this barter, which relies on Andean social traditions of reciprocity, gives campesinos stability when facing the uncertainty of year-to-year variation in farm production, fluctuation of currency values, and changing market prices.[1] Overall long-term security is more important than short-term profit. Nonmonetary exchange is thus interpreted both as resistance to the market as well as recourse to a traditional moral economy. According to Mayer (2002: 137), because such reciprocal relationships offer these specific advantages, they will remain important to Andean livelihoods in spite of increasing market integration.

Here Mayer (2002: 143) is drawing on an anthropological theory of barter that understands it to be only "one mode of exchange amongst others" (Humphrey and Hugh-Jones 1992: 6). It can be used in situations with "an absence of money and where there is no over-arching monetary system, but also where a common currency exists but where people prefer not to use it, or where there is not enough money to go round" (Humphrey and Hugh-Jones 1992: 4). Barter usually occurs in face-to-face situations among individuals of equal status (at least in principle). In addition, since it is almost always a repeated transaction, exchange equivalencies become standardized and there is an obligation to act fairly (Humphrey and Hugh-Jones 1992: 6–8). In this sense, nonmonetary exchange, while perhaps representing only a small percentage of total exchanges, is shown to be a reasonable response to issues encountered in daily life. It comprises a small, but critical component of highland livelihoods (Mayer 2002: 169).

Given this interpretation of barter, it is important to remember that even when goods are exchanged without the use of currency, bargaining, or haggling is still used to adjust rates of exchange. Equivalencies are often achieved through manipulation of characteristics other than price, such as the method of measurement used and the quality of products traded (Kula 1986: 102–9; Polanyi 1975: 150).

[1]For a more historical perspective, Larson (1995: 24) similarly explains that in the colonial period, "such non commodity forms of circulation continued to provide varying degrees of social insurance—and alternative commitments—against the vagaries of the market, uncertainties of climate, and extractive pressures of the state."

These characteristics are in turn affected by a variety of features, including location of trade, seasonality, and relationships between traders (Kula 1986: 109). Even if the *price* in an exchange remains constant, as in Mayer's (1971) classic, although fictitious, example of one sheep for one sack of potatoes, the quality of the tubers, the plumpness of the sheep, and the volume of the sack are of obvious import to the participants in exchange. Here, just as Turner and Williams (2002) discuss in their description of "real markets," (see Introduction), it is important to look closely at pricing systems and exchange rates. This is especially critical when value is not expressed as an explicit monetary amount, and when price is controlled by factors that are not strictly economic. These factors may have much to do with the sociocultural context of exchange, as in Humphrey and Hugh-Jones' (1992) description of barter systems, and Kula's (1986) analysis of measurement systems and social relations.

This discussion of barter does not mean to imply that Andean people participate exclusively in only one or the other of these two spheres of exchange. *Campesino* households take advantages of opportunities in both, moving within and combining the capitalist market and "traditional" structures of exchange (Göbel 1998: 868). This is by no means a recent development. As Larson (1995: 21) demonstrates, throughout both the colonial and post-colonial periods Andean social groups struggled to combine nonmonetary reciprocal relationships with tribute and mercantile systems.[2] One important result of this merging of practices was that subsistence needs often came to be met through the use of "…strategies involv[ing] interethnic exchange that etched new patterns of circuitry and alliance across the Andean landscape, quite apart from the flow of commodities between the Andean and Spanish sectors" (Larson 1995: 22). These patterns could be something akin to Murra's (2002 [1972]) vertical archipelagos; but after the sixteenth century, barter and other trade was increasingly used for transferring goods between ecological zones (Larson 1995: 22; Assadourian 1995). Fonseca (1973) likewise emphasizes that methods of trade as well as participants in different types of exchange have varied with time, and that the practices observed now are not necessarily representative of practices from other eras.

The historical contingency of these practices, while undeniably important, is difficult to generalize. Certainly, there is great geographic variability in the ways that Andean people interacted with and adapted to the major economic systems discussed in the works listed above (e.g., tribute, mercantilism, modern capitalism). Such a reality is currently being observed in geographic studies of the locally specific effects of neoliberalism (McCarthy 2006). All of this evidence indicates the importance of place-based analyses. Returning to the Olleros case study, recent large-scale events, especially Peru's Agrarian Reform Law of 1969 (and its lead-up), have had a large impact on potter livelihoods. The specific implications of

[2]See also Stern (1995) for an analysis of how Andean peoples combined their "traditional" economy with the European market economy starting from the earliest days of the Spanish conquest. Additionally, see Larson (1998) for a more detailed description of the evolution of barter systems in Cochabamba, Bolivia.

this process for Olleros pottery exchange practices are discussed in the following chapters.

When taken as a whole, the Andean Studies literature clearly shows that individual campesino households, especially in the rural highlands, work within a flexible framework of exchange. Within this framework, they try to supply themselves with food and other necessities from a variety of ecological zones. To accomplish this, they use a variety of mechanisms for transferring goods and labor, which likely include both market and nonmarket components. The ways these activities are pursued are likely to be locally unique even to the household level, and contingent upon specific conditions and histories. From this broad perspective, the Andean landscape may be understood as a landscape of exchange, with spatial patterns resulting from the physical environment, the distribution of productive activities, and social and economic organization.

1.3 The View from Development Studies and the "New Rurality"

A different perspective on exchange and rural livelihoods comes from work in the field of Development Studies, and especially the work with a focus on livelihood diversification and the so-called "New Rurality" (Kay 2008). This literature suggests ways to link the Andean situation with the broader context of rural livelihoods in Latin America and the Global South more generally. Themes of particular importance include diversified livelihood strategies, nonagricultural/off-farm income activities, and the balance of subsistence production with market integration.

Diversification is one way to conceptualize the varied agricultural and nonagricultural activities performed by individuals and households as they attempt to earn a living. In the Andes, as in many other rural areas, diversified livelihood strategies include activities such as farming, livestock raising, (migratory) wage labor, and entrepreneurial activities. Diversification likely entails engagement with both the subsistence and market spheres, as well as the spatial distribution of activities across multiple locations (Valdivia et al. 1996). Several explanations have been presented for why households would choose to diversify instead of specialize: risk management and smoothing of income throughout the year or across predictable seasonal cycles are just a few (Reardon et al. 1992; Valdivia et al. 1996). Diversification has also been seen as a response to poverty (Reardon and Vosti 1995) and to declining investment in agriculture. For example, Swinton and Quiroz (2003) found that families in the Peruvian altiplano preferred to support (seasonal or semi-permanent) emigration of family members to rainforest or urban locations where employment is possible, instead of focusing attention on agricultural intensification or conservation (see also Gray 2009 on migration streams). The implications of a high level of diversification are complex; within this framework livelihoods may become more stabilized, but "hyper" or "distress" diversification can also result in increased (or sustained) levels of poverty (Rigg 2006: 194, see also Kay 2008).

Pottery production is just one component of a diversified livelihood strategy, falling under the category "rural nonfarm income activity." Nonfarm activities constitute a large component of rural livelihoods, and their proliferation has been seen as characteristic of the "New Rurality" and increasing involvement in a neoliberal market (Kay 2008; see also Zimmerer 2014). In Peru, Escobal (2001: 497) estimated that as much as 51% of rural income is earned from activities outside of own-farming. Here, nonfarm refers to activities outside of agriculture, in either manufacturing or services. According to Reardon et al. (2001: 396), rural households are motivated to pursue nonfarm activities by "pull factors" (e.g., earning higher incomes) as well as "push factors" (e.g., risky farming, land constraints, lack of insurance/credit). As in diversification, nonfarm activities are used for many reasons, including to smooth inter/intra-year variation in income, to increase income or alleviate poverty, to manage risk and cope with income shocks, and to finance investments in farm or other assets (ibid). Reardon et al. (2001) also found households with poor agricultural potential tend to rely heavily on nonfarm activities. However, while rural-based nonfarm income comprises a large share of the total income of such poor households, the amounts they actually earn are relatively low compared to their wealthier peers. It follows that households pushed into nonfarm activity to survive generally participate in "low-return, low productivity 'refuge' employment" with few "poverty-alleviation effects" (Reardon et al. 2001: 403, see also Elbers and Lanjouw 2001). In reality, these activities should be seen as a means of survival instead of a means of profit or advance (Kay 2008).

De-agrarianization, or a decline in rural peoples' commitment to farming, has been one outcome of increasing diversification and reliance on nonfarm income (Rigg 2006). While this trend is not necessarily problematic, as Reardon et al. (2001) imply, it is important to consider the terms at which this income is earned. Rigg's (2006) conclusion is that while in the past poverty was often said to have been caused by remoteness and lack of market integration, and now poverty may be more accurately explained as the result of development processes, which actually is through engagement with the market and the state that rural people become (or remain) poor. Migration and remittance incomes have been found to have diverse impacts on rural livelihood activities, especially in the Andes (for example, Bebbington 2000; Jokisch 2002; Zimmerer 2014).

This discussion points to broader issues of peasant economics and how peasants interact with the market economy. It is well known that peasant livelihood strategies are usually neither purely subsistence nor purely market driven, but exist along a continuum between these two extremes (Deere and de Janvry 1979: 602; Valdivia et al. 1996). Eric Wolf noted this a half century ago:

> The peasant is not engaged in agriculture alone…we must first ask questions regarding either the degree to which each peasant household carries on the necessary craft specialties or—correspondingly—the degree to which these specialties are in the hand of another whom he must pay in food for their specific services (Wolf 1966: 37).

Thus a key feature of peasant economic strategies is, and perhaps has always been, that they allow for simultaneous, and to a certain extent selective, articulation with both the market and subsistence spheres. That said, increasing integration in a neoliberal economy has generally been linked with a decrease in access to resources previously used for production, including land, and thus possibly an increase in diversification and nonfarm activities (Kay 2008). In the Andes, Göbel (1998: 868) interpreted this tendency as both an active risk-spreading strategy, but also as a reaction to the limited opportunities available to small farmers, herders, and artisans. Commodity chain analysis, with a focus on production, distribution, and consumption of goods, can provide a frame for understanding this selective market integration (Turner 2007). In general, a definition of "rural livelihoods as dynamic and diverse rather than static and based on a single economic activity" (Perreault 2009: 450) provides the most realistic model for understanding both selective market integration and livelihood diversification strategies.

1.4 Andean Pottery Production and Trade

Many works about Andean economies and exchange systems briefly mention the exchange of clay pots (Brush 1980: 221; Delgado and Ponce 2002: 7; Flores Ochoa 1977: 146; Fonseca 1973: 123; Gade 1975: 52; Karasik 1984: 58; Mayer 2002: 150; Orlove 1974: 294). Yet, these works pay more attention to the exchange of other products, specifically agricultural products. This may be because these scholars' interests lie in linking ecology and economic organization, especially in relation to Murra's concept of verticality. Thus trade of manufactured objects is not as well understood as trade of agricultural products, and pottery distribution and exchange has not often been given deep consideration (exceptions include Mohr 1992; Sillar 2000; Ramón and Bell 2013; see also Livingstone Smith 2016 for Central Africa). After all, pottery can be produced in a range of ecological zones, assuming the availability of appropriate clay sources. It may not, therefore, provide great insight into ecological complementarity.[3] Yet, as will be outlined below, pottery production and exchange are inextricably linked to Andean agropastoral organization, thus providing a unique vantage point from which to examine the links between handicraft production, agriculture, and rural exchange practices.

Highland pottery production, like much craft specialization in the Andes, is almost always a secondary activity performed in conjunction with agriculture or livestock raising (Fonseca 1973: 156; Mohr 1992: 68; Sillar 2000: 49; Ramón 2013a, b; Ramón and Bell 2013). Often the quantity of pots produced is related to how much time is devoted to these primary activities, as well as how much of the household income is earned from these pursuits (Mohr 1992: 68). Sabogal (1982: 41), in his

[3]See Mohr (1992) for an attempt at discussing pottery production with respect to complementarity, see Ramón and Bell (2013) for an analysis of pottery, verticality, and archaeology.

extensive survey of the potters of Piura, (Northern Peru), found that highland pottery production always complements farming. But he also found that it often goes hand in hand with land scarcity and/or land of poor agricultural quality. Pottery is a skill or capacity a household may draw upon to supplement its other activities. It may be especially useful if income from those other activities is limited.

Thus pottery corresponds with Reardon et al.'s (2001) assessment of one kind of nonfarm activity (the "Z-good" category defined by Hymer and Resnick 1969), which occurs

> …at home and on the farm rather than off-farm in the village or town, in small-scale operations, using traditional technologies, intensive in labor and home-consumed or sold to the local market. These are goods for which the demand does not increase and the technologies used do not improve/evolve. Labor supply is driven by push factors, and labor demand is local, low level, seasonal, and fueled by semi-subsistence farm incomes… (Reardon et al. 2001: 397).

In other words, it is a nonfarm income activity that requires little input of capital but that is highly labor intensive. Additionally, as we will see, demand for the product is generally not increasing, although pottery exchange continues to be a viable livelihood strategy. Especially in Olleros, potters do not need cash to make pots. Since pots are mainly bartered for food crops with farmers, they may be transformed into food through trade. One result of these barter practices is that potters do not need money or farmland to supply their households with (at least some) food. Pottery production thus becomes a useful strategy for households with limited access to other resources, including land and labor. Not unexpectedly, the production and use of clay pots has often been deemed an indication of poverty, and sometimes carries a connotation of being a *runa*, or indigenous activity (Orlove 1998). This book explores the relationship between poverty and pottery production and makes the argument that pottery is not the direct result of poverty, but that it is a useful skill for poor households. This distinction is more fully developed within the following chapters (see especially Chap. 4).

1.5 Pottery Production in Piura

The Department of Piura, located in the extreme north of Peru, has long been recognized for pottery-making. After his extensive survey, found that

> One of the most astonishing things to the traveler is the exceptional number of ceramicists in the region of Piura. If one is deeply curious, he will continue to discover more and more potters in the interior. The first impression is of the insatiable necessity of ceramics in the indigenous campesino culture (Sabogal 1982: 19, my translation).

While the production of *tinajones*, which are large jars for brewing *chicha* (maize beer), in Piura's coastal town of Simbilá is well known and well studied (Camino 1982), the production of pottery in the Piuran highlands has only recently become the focus of in-depth analysis (Ramón 2008; Ramón and Bell 2013, although see the aforementioned Sabogal 1982). Even though the Piuran sierra is relatively low in elevation, with the highest altitudes not quite reaching 4000 m.a.s.l., highland pottery production follows a very different pattern than the coastal example. There are four communities in the highlands where pottery is currently produced: Santo Domingo, San Bartolomé de los Olleros, San Juan Bautista de Sondorillo, and Santa Ana de Huarmaca (Ramón 2008; Ramón and Bell 2013; see also Monzón 1991) (see Fig. 1.1). In these communities potters produce utilitarian clay vessels for domestic use, especially cooking. The production is at the household scale, with nothing approaching the large-scale workshops in the coastal centers. Consequently, the final products are different. Highland potters make a range of vessel forms; the most important of which are the plain cooking pot (*olla*) and the toasting pot used mainly for toasting maize and wheat (*tiesto de tostar*) (Fig. 1.2).

The four highland communities can be divided into two distinct groups based on production technique: Santo Domingo and Olleros fall into one group, Sondorillo and Huarmaca fall into the other (Ramón 2008). For the purpose of this paper, the most important patterns of pottery production within these two groups are: (1) producers in Santo Domingo and Olleros are all women, while in Sondorillo and Huarmaca men are the potters; and (2) production in Olleros and Sondorillo is much more intense than in Santo Domingo and Huarmaca, where few individuals still participate in this craft.

This pottery exchange project grew out of a much larger ethnoarchaeological study of the pottery production of the Northern Peruvian Andes conducted by Ramón (2008, 2011, 2013a, b) and Ramón and Bell (2013). In the course of this

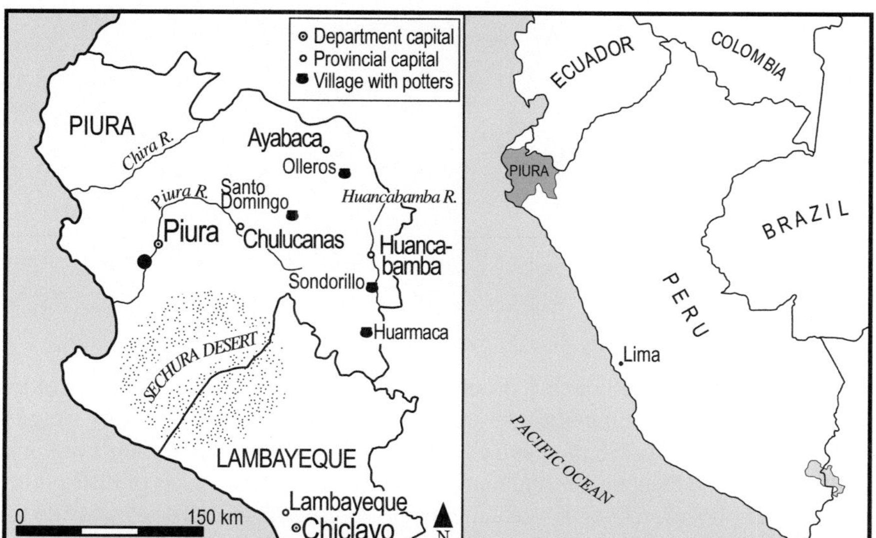

Fig. 1.1 Study location and map of villages with potters in Piura, Peru (Map by author)

Fig. 1.2 Main pot forms made in Olleros. All pots are in production and unfired. *Top* Regular cooking pot (*olla*), being made by Orfilia Mondragón. *Middle* Toasting pot (*tiesto*), being made by Hortensia Criollo, *Bottom* back row: regular pots with necks (*ollas*), front row: pots without necks and with handles (*sapimas*), being made by Micaela Jiménez (Photos by G. Ramón)

project Ramón studied 30 pottery production communities. I traveled as part of his research team to 11 communities, between July–November 2004 and July–August 2005. While pottery distribution and exchange were addressed in Ramón's regional study of production techniques and material culture, detailed analysis of distribution methods and potter livelihoods was not a primary research goal. To select a site for this study, I narrowed the list of the places visited during the 2004–5 fieldwork to two choices: Sondorillo and Olleros. These were the two places where pottery production and trade were most abundant. They were also interesting because of their extreme northern location, and therefore provided complements to previous analyses of pottery production and exchange in Southern Peru and Bolivia (Sillar 2000; Mohr 1992), as well as Andean exchange studies more generally, where the focus has most strongly been placed on the Central Andes (c.f. Fonseca 1973; Mayer 2002; Mayer and Alberti 1974).[4] In the end, I focused on the single site of Olleros.

1.6 Pottery Production in San Bartolomé de los Olleros

San Bartolomé de los Olleros *campesina* community of approximately 350 members (*comuneros*) and their children. It is located approximately five hours walking or four hours by car from the highland city of Ayabaca, which is the provincial capital and the site of an important annual Catholic pilgrimage (*Poderoso Señor Cautivo de Ayabaca*). As the name Olleros, ("potters" in Spanish), suggests, potters have long been present in this community, at least since the colonial period and possibly since pre-Hispanic times. The oldest reference to the "Hacienda Olleros" dates to 1621; another document from 1645 also alluded to the presence of potters at the site (Ramón 2008: 495).

Today's Olleros is divided into six sectors (called *caseríos*): Aguayco, Congolí, Sidro, Toronche, Cafetal, and La Pampa (Fig. 1.3). Pottery is currently produced in the last two (Cafetal and La Pampa). These are also the lowest in elevation, at 1411 and 1541 m.a.s.l., respectively. These sectors fall in the transition zone from "yunga" to "quichua" (Bernex and Revesz 1988), and within Tosi's (1960) "subtropical dry forest" life zone (*bosque seco subtropical,* bs-St). They are fairly recent settlements, formed after the Agrarian Reform broke apart the Hacienda Olleros in 1972. During hacienda times, pottery was produced in the sectors of Aguayco and El Sidro, where many families still maintain fields. After the Agrarian Reform, almost all potters (and other residents) of these sectors migrated to the lower elevations of the community's territory, founding the sectors of La Pampa and Cafetal (see Figs. 1.4 and 1.5). The community of Olleros is situated within a mosaic of settled farming communities which grow a range of products (e.g., coffee, sugarcane, maize, wheat, peas). Many of these communities are ex-haciendas, and many of them are home to pottery consumers.

[4]Many of the observations made during studies of exchange are related to agricultural production and the idea that ecological complementarity is a motivating factor for exchange. Looking at trade practices in a different environmental context can only work to expand our understanding of Andean exchange practices. The Northern Andes, and Piura specifically, differ from the Central/Southern Andes in several important ways: especially in ecological zones (*páramo* instead of *puna* at high altitudes), and climate patterns (warmer temperatures, more rainfall) (see Brush 1977 or Salomon 1986 for a fuller treatment of these topics). Since these features so strongly influence agricultural practices, it is expected that they will also affect exchange practices.

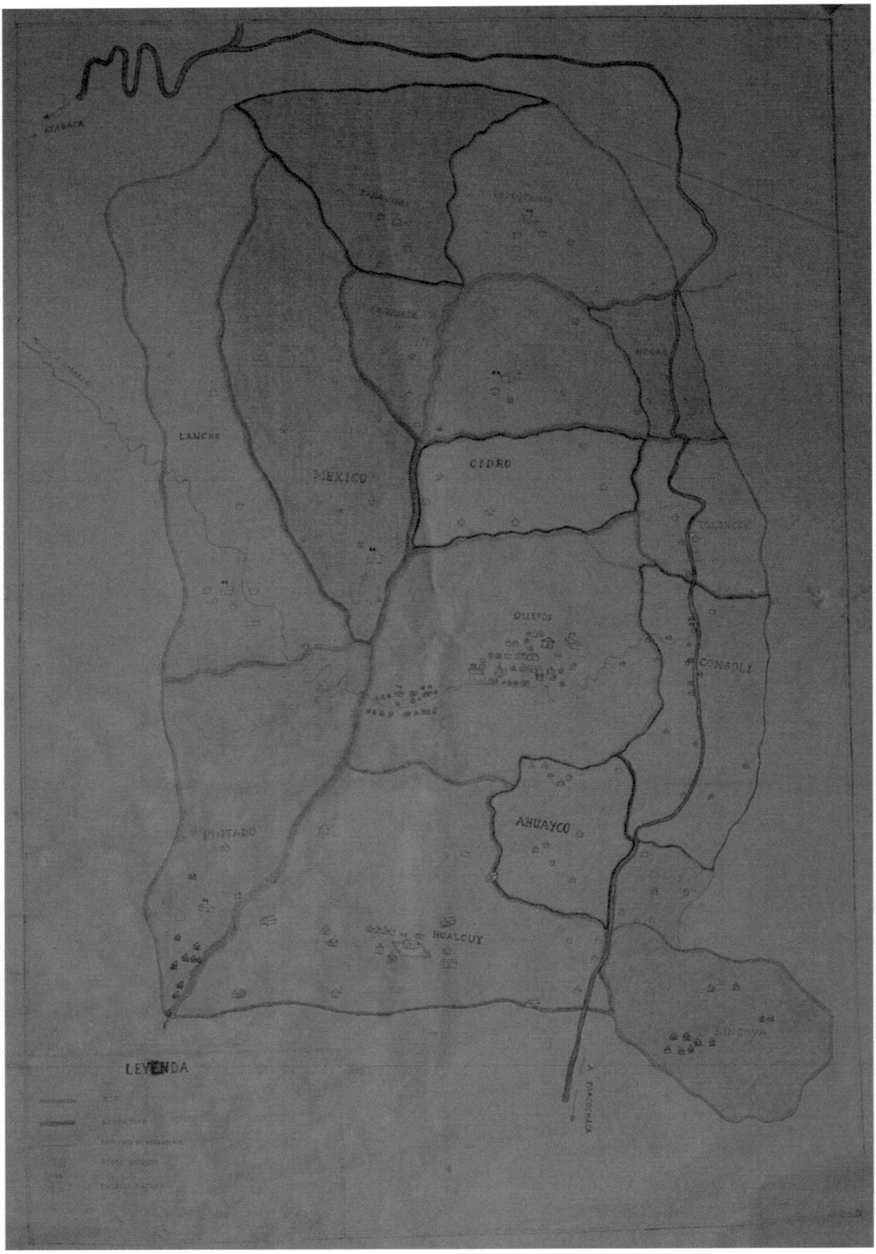

Fig. 1.3 Hand-drawn map of the sectors (*caserios*) of Olleros and the sectors of the communities immediately surrounding it. The map was created by the nurse at the Olleros health center (*posta de salud*). Olleros sectors displayed Aguayco, Congolí, El Sidro, Toronche, Cafetal, and La Pampa (Photo by author)

Fig. 1.4 View of Olleros sector La Pampa, from the footpath connecting La Pampa with the city of Ayabaca (Photo by author)

In Olleros, pottery is a female activity. Many of the resident women (*comuneras*) know how to make pots, even though not all actively practice the craft. The technique used to produce the vessels is fully described in Ramón (2008), but several of the most important aspects of this activity are described here. Clay is currently collected from one mine, which is conveniently located between the two pottery producing sectors of the community. Potting is done entirely by hand, using simple tools: potter's plates, large wooden "knives" (*caiguas/callhuas*), wooden polishing sticks, pieces of gourd, a metal knife, leaves from a chirimoyo or coffee tree, and a wooden decorating tool (Fig. 1.6). Potters work alone or in small work groups (see Sect. 2.6). Work is done outside in shaded areas near the potter's house. It is a dry season activity practiced between the months of May–December (with a peak in July–August). This is due to firing requirements (e.g., dry fuel, dry ground, no rain during firing), as well as to the local agricultural calendar (more detail below). Potters make a variety of vessel forms, each of which may also come in a variety of sizes. They are regular cooking pots (*ollas*), neckless cooking pots (*sapimas*), toasting pots (*tiestos*), frying pans (*cazuelas*), and water/chicha jars (*cántaros*, *tinajas*) (see Fig. 1.2). Pottery is an activity that does not require monetary investment: the tools used are usually produced from local sources, the clay is available to community members without charge, and pots are fired mainly with firewood or cow dung which is also free (Fig. 1.7). These pots can then be traded by a variety of methods for both food and money.

Fig. 1.5 View of Olleros sector Cafetal, from above, looking in the direction of the city of Ayabaca (Photo by G. Ramón)

Fig. 1.6 Typical pottery-making toolkit from Olleros (Photo by G. Ramón)

Fig. 1.7 Micaela Jiménez collecting clay from a local community source (Photo by G. Ramón)

In general, girls learn to make pots as teenagers, by watching the work of their mothers, aunts, and neighbors. It takes about two years to fully learn the craft and to produce pots of satisfactory quality for exchange. Additionally, many young women from other places learn from their mothers-in-law after marriage brings them to Olleros. Nowadays, the craft seems to be passed down less frequently than in past generations; only about half of the potters interviewed had already taught their daughters (5 of 11). Of these daughters, not all practice the craft. See Appendix A for profiles of the potters and traders interviewed.

Pottery production is never the only activity pursued by a household. Every potter household also has agricultural responsibilities (see Table 1.1). The major crops grown by potter households are maize, manioc, and plantains for household consumption; peanuts for sale to merchants; and sugarcane for production of

Table 1.1 Annual cycle of agriculture and pottery in Olleros

Crop/activity	Maize “hibrido blanco”	Peanuts “rojo” or “negro”	Year-round Tropical crops	Pots
Jan	Plant	Plant	Plantain: Plant any time of the year, bears fruit in 2–3 years	
Feb		Plant		
Mar		Plant	Manioc: Plant in March, Harvest in 1 year	Begin potting season
Apr	Choclo Harvest			Potting
May				Potting
June	Harvest	Harvest		Potting
July	Harvest			High potting season
Aug				High potting season
Sept				Potting
Oct				Potting
Nov				Potting
Dec	Plant when rains start		Sugarcane: Plant in rainy season, mature in 2 years	End potting season

Rains occur between December and April (especially January–March). “Summer” falls between May and October

Fig. 1.8 *Chancaca* production in La Pampa by the Flores family. Currently, there are no potters in this family (Photo by G. Ramón)

chancaca ("*dulce*") and liquor (*cañazo*), which are both bartered and sold[5] (Fig. 1.8). In both sectors agriculture is a difficult endeavor, as one potter stated "Yes, one can plant, it's just that…they [the fields] don't carry…" (Emilia Pintado).[6] The potting season (May–December, with peaks in July–August)

[5]*Chancaca* (also known in Spanish as *raspadura*) is a kind of hard brown sugar from which the molasses has not been separated. Córdova (1990: 87) describes the production and trade of this product well: "A common product exchanged is '*la chancaca*' or '*dulce*'. This is a product of sugarcane, extracted using a '*trapiche*' (a milling device)…. All of the juice is extracted…boiled until it is viscous… it is then put into small wooden molds specially made to shape 'loaves' of sugar. When cooled, two loaves are put together and wrapped in the dry banana leaves. These '*atados*' (2 loaves together) are then ready for market or barter. This product is taken to the *yunga* and the *quechua baja* to '*Los Altos*' [among other places] where it is traded for oca, wheat, and wool" (my paraphrase and translation). Weismantel's (1988: 109) description of "*rapadura*" or "*panela*" slightly adds to Córdova's description; in Zumbagua, her study site in the Ecuadorian highlands, "*Rapadura* is more expensive than white sugar and is therefore disappearing from the diets of most households, except as an occasional luxury." In Olleros, *chancaca* is often traded alongside pots, but it is more sought after than pots among the customers/traders who visit Olleros, and it is arguably of higher prestige.

[6]"Si se puede cultivar, si que no, es que… no cargan igual pues" (Emilia Pintado). A note on annotation: English translations of quotations taken from interviews are included in the text. The Spanish transcriptions are provided as footnotes, since sometimes multiple interpretations of respondents' words are possible. Respondents' full names are used, based on discussion with potters who both deserve credit for their opinions and appreciate the publicity for their work (and any new clients it may bring).

corresponds with the agricultural cycle, it begins after the maize/peanut harvest when there is a lag in farming responsibilities. It is important to note that different households have different land holdings and different commitments to each of these crops (i.e., some households have no sugarcane, some rent land instead of owning it). This information was alluded to in many of the interviews with potters, but proved to be uncomfortable topics for some. Consequently, differentiation among potter households is not dealt with systematically in this study.

Many households also raise livestock. Most families own pigs and fowl, (chickens, ducks, and turkeys), for sale and consumption; many own donkeys or horses for transportation; and some own bulls for powering the sugarcane mills (called *trapiches*). Most women weave textiles, using a backstrap loom, for their own use or as piece-work in peonage relationships. Additionally, many households have members who migrate to urban centers or to the coffee and coca-growing zones in the east to work. Again, the importance of these other occupations varies among households. Pottery production is integrated into this range of livelihood activities. Given this generalized household economic portfolio of resources, and considering basic Andean culinary standards, it is clear that much of food consumed by households in these villages must be brought in from somewhere else. Trading pots for food, especially grains, is one way of resolving this situation.

1.7 Research Methods

This is a qualitative study of livelihoods, with a specific focus on one single activity. The major research methods used were semi-structured interviews and (participant) observation of daily activities. The interviews collected from this fieldwork were transcribed and analyzed using narrative analysis techniques as well as cartographic visualization and spatial analysis methods.

1.7.1 Data Collection

Sixteen individuals in Olleros were interviewed from the period June 15 to June 26, 2006, with the assistance of G. Ramón. These interviews took the form of semi-structured conversations based on a list of topics developed prior to the fieldwork period (see Appendix B). This list was derived from prior experiences among potters (2004–5), and especially three previous research visits to the community of Olleros in 2004. Parts of this list also came from review of the literature, most specifically by the methods employed by Sillar (2000). Additionally, other important topics emerged during the course of the conversations with potters and pottery traders; these topics were incorporated into subsequent interviews when appropriate. The style of questioning used was based on the "grand-tour" questions described by Spradley (1979) (e.g., general description of barter trips), which were followed up when possible and appropriate by "mini-tour" questions about specific exchange experiences (e.g., description of the most recent barter trip), as well as other queries for more detailed explanations. Overall, the goal was to elicit descriptions of what pottery distribution methods are used, where exchange occurs, how trading partners are found, what products are exchanged, how transactions are carried out, and how these practices might be changing with time.

The sixteen individuals interviewed were potters, pottery traders, and other residents of Olleros. Multiple interviews were conducted with almost all of these individuals (see Appendix A for profiles of respondents). I was already acquainted with many potters from previous stays in Olleros, so these individuals were approached first. To recruit additional respondents, a network sampling strategy was used: the first potters interviewed were asked to introduce us to others who would be interested in participating, and so on. Interviews generally took the form of the researchers visiting the home of the individual, where we spoke with the potter/trader, but also with family members who were around and willing to talk. All interviews were conducted in Spanish, which is the native language of Olleros residents (Quechua is not spoken in this part of Peru), and were recorded with the permission of the interviewees. To the extent possible these recorded interviews were transcribed during the evenings in the field so that specific follow-up questions could be formulated.

While this interview style was effective for gaining information about the specific topics included in the initial questionnaire, there were limitations to the approach which warrant recognition. First, the interview style heavily tilted the conversations toward the topics on the list, perhaps preventing others from arising, and in some cases (unfortunately) providing a structure which made it difficult for subjects to discuss related topics that were not specifically queried. A second limitation was that the population interviewed was biased. Almost all of the active potters in the community were interviewed, although a few were not contacted or chose not to speak with us; but we spoke with fewer individuals who trade but do not produce pots (this includes husbands of potters who are responsible for most of the trading trips); and we contacted even fewer of the other residents who do not make but who may trade pots. A third limitation has to do with the completeness of the information collected. The list of places of pottery distribution presented here is likely not complete, for any number of reasons the list is partial. Still, without a doubt, the description does capture the most important trading sites, and is an accurate depiction of how Olleros pots are traded. Finally, it was only possible to remain in the community for a short time period, which limited the total amount of information collected.

The second component of the fieldwork was (participant) observation. The proposed research planned a significant component of participant observation accompanying pottery traders on trading trips as well as observing exchange practices occuring within the community. However, due to personal injury, walking long distances became impossible, and so unfortunately the traveling component had to be cut out of the fieldwork. My inability to walk long distances may also have influenced which individuals were interviewed; as it turned out, our nearest neighbors were interviewed most extensively. Additionally, the fieldwork period fell slightly before the peak trading season began, and only a few trading trips took place during the time we were there.

Finally, it is important to be reflexive about our position in the community as researchers. As this was our fourth visit to the community, many residents already knew and trusted us. Our position was further strengthened by our friendship with Daniel Mondragón, the President of the Ronda Campesina (an important local authority: the leader of the community's civilian defense organization). Describing the research to him was important for making our purpose publicly understood. However, none of this completely worked to overcome the drawbacks of being "gringo" university students (thus urbanites) visiting a rural highland community. Surely this position limited what we were able to see/understand. Finally, in some cases it was easy to sense the discomfort that individuals felt when describing realities of poverty, land shortages, racial tensions, and other personal aspects. Almost all of these issues were anticipated entering into the project, and while they limited the depth of analysis possible, they by no means prevented the development of an understanding of pottery and livelihoods.

1.7.2 Data Analysis

Two analytic techniques were applied to the data collected during the fieldwork period. The first was narrative analysis of the interview data and the second was cartographic representation of the spatial data relating to pottery exchange. The following paragraphs describe each of these methods.

All interviews were digitally recorded with the permission of the respondents. These interviews were transcribed verbatim. Narrative analysis methods were then applied to the interview data. Specifically, the transcriptions were coded for a number of concepts and themes. This list was derived from the initial questionnaire as well as from the content of the responses. These topics included specific details ("concepts") related to the following broader "themes:" trading trip destinations, customer points of origin, community-based exchange, peonage, products traded, potter life cycles, seasonal agricultural and trading calendars, chancaca production, pottery as a livelihood strategy, and change with time.

To code the interviews, specific descriptions of each of the topics on the list were identified and compared. For example, when trying to understand how customer visits have changed with time, different potters' opinions were read carefully and compared with one another. Statements about this situation include

> …in the time when we were making pots there in Aguayco [the sector of Olleros where people lived during the hacienda], more people came on Sundays to buy pots (Irma Abad).

> Well, now since....now it's been thirty years more or less, because after that the people stopped coming down here [to La Pampa, the sector of Olleros where people live now] (Luzmila Parihuamán).

> now [pots] aren't made, it's because now there isn't any market... (Oralia Mondragón).[7]

Analysis of the remarks made by the three potters Irma Abad, Luzmila Parihuamán, and Oralia Mondragón were read, placed in the context of their specific conversations, and combined to form the analysis of customer visits presented in Sect. 2.4. A similar process was carried out for each concept and theme. Details from each response were evaluated and woven together to form the descriptions presented here. These descriptions tried to explain general, -wide trends, while at the same time incorporating differences among the responses, in both factual details as well as perceptions of larger issues. Through analysis of each of the coded concepts and themes, the depiction of pottery exchange presented here emerged.

Given the spatial component of the theoretical basis for this study, especially the multi-community approach, mapping the exchange networks added important depth to the data analysis. These network maps generally took the form of lines radiating outwards from Olleros toward the points of distribution, rather than the form of a bounded territory of distribution. This follows the method demonstrated by Smith (2005); it was useful for highlighting the spatial patterning of specific features related to exchange.

The maps were created using data from diverse sources. Basemap data was acquired from the geo-server of Peru's Ministry of the Environment (district, province and department borders), from ESCALE (online statistics center) of the Ministry of Education (locations of villages and towns, topographic data, and rivers) and from the Ministry of Transport and Communications (local, district, and national road networks). The digital elevation model (DEM) was constructed with ArcGIS 3D Analyst Tools, using the previously mentioned topographic data. The final maps were made using ArcGIS and Adobe Illustrator. The locations (and spellings) of distribution sites mentioned by the potters and traders follow the ESCALE data. However, as anyone who has spent time in the Andes will understand, locating a "community" as a single point on a map greatly simplifies the reality that each community often encompasses a wide swathe of territory. So these "points" should be interpreted only as close approximations to where pottery trade actually occurs. The "routes" drawn on the various maps are approximations based on potter and trader descriptions, and on topographic realities; they should not be understood as exact representations of the paths followed. Chapters 2 and 3 present these maps and provide description and analysis of the features they illustrate.

[7]Opinions include "...en tiempo que hacíamos ollas allá en Aguayco [sector of Olleros where people lived during the hacienda], el día domingo venía más gente a comprar ollas." (Irma Abad); "bueno, ya hace...ya como treinta años más o menos porque de allí esa gente ya dejó de venir acá abajo..." (Luzmila Parihuamán) and "ya ahorita ya no se hace, es por el motivo que ya no hay negocio...." (Oralia Mondragón).

References

Assadourian CS (1995) Exchange in the ethnic territories between 1530 and 1567: The *Visitas* of Huánuco and Chucuito. In: Larson B, Harris O, Tandeter E (eds) Ethnicity, markets, and migration in the Andes. Duke University Press, Durham, pp 101–134

Bebbington A (2000) Reencountering development: livelihood transitions and place transformations in the Andes. Ann Assoc Am Geogr 90(3):495–520

Bernex N, Revesz B (1988) Atlas regional de Piura. CIPCA and PUCP, Piura

Brush SB (1977) Mountain, field, and family: The economy and human ecology of an Andean valley. University of Pennsylvania Press, Philadelphia

Brush SB (1980) Peru's invisible migrants: a case study of inter-Andean migration. In: Orlove BS, Custred G (eds) Land and power in Latin America. Holmes & Meier Publishers Inc, New York, pp 211–229

Camino L (1982) Los que vencieron al tiempo. CIPCA, Piura

Córdova H (1990) Tecnologías en el sistema de producción del departamento de Piura. In: Bernex N, Córdova H, Roux JC (eds) El proyecto de desarrollo rural integral de la sierra central del departamento de Piura (Convenio PUC-ORSTOM): Tercera parte-aspectos socio-económicos. PUCP, Lima, pp 78–87

Deere CD, de Janvry A (1979) A conceptual framework for the empirical analysis of peasants. Am J Agric Econ 61(4):601–611

Delgado F, Ponce D (2002) La Feria: Fiesta de Siete Viernes. Revista Compas 5:7–9

Elbers C, Lanjouw P (2001) Intersectoral transfer, growth and inequality in rural Ecuador. World Dev 29(3):481–496

Escobal J (2001) The determinants of nonfarm income diversification in rural Peru. World Dev 29(3):497–508

Flores Ochoa J (1977) Pastoreo, tejido e intercambio. In: Flores Ochoa J (ed) Pastores de puna, Uywaminchiq punarunakuna. IEP, Lima, pp 133–154

Fonseca C (1973) Sistemas económicos andinos. Biblioteca Andina, Lima

Gade D (1975) Plants, man and the land in the Vilcanota valley of Peru. Dr. W. Junk B.V. Publishers, The Hague

Göbel B (1998) 'Salir de viaje' Producción pastoril e intercambio económico en el noroeste argentino. In: Dedenbach-Salazar Saenz S, Arellano Hoffmann, C, Köning E, Prümers H (eds) 50 años de estudios americanistas en la universidad de Bonn—Nuevas contribuciones a la arqueología, etnohistoria, etnolingüística y etnográfica de las Américas, vol 30. BAS, Bonn, pp 867–891

Gray CL (2009) Environment, land, and rural out-migration in the southern Ecuadorian Andes. World Dev 37(2):457–468

Humphrey C, Hugh-Jones S (1992) Introduction: barter, exchange and value. In: Humphrey C, Hugh-Jones S (eds) Barter, exchange and value: an anthropological approach. Cambridge University Press, Cambridge, pp 1–20

Hymer S, Resnick S (1969) A model of an agrarian economy. Am Econ Rev 59(4):493–506

Jokisch BD (2002) Migration and agricultural change: the case of smallholder agriculture in highland Ecuador. Hum Ecol 30:523–550

Karasik G (1984) Intercambio tradicional en la puna Jujeña. RUNA, XIV, pp 51–91

Kay C (2008) Reflections on Latin American rural studies in the neoliberal globalization period: a new rurality? Dev Change 39(6):915–943

Kula W (1986) Measures and men. Princeton University Press, Princeton

Larson B (1995) Andean communities, political cultures, and markets: The changing contours of a field. In: Larson B, Harris O, Tandeter E (eds) Ethnicity, markets, and migration in the Andes. Duke University Press, Durham, pp 5–53

Larson B (1998) Cochabamba, 1550–1900. Colonialism and agrarian transformation in Bolivia. Duke University Press, Durham

Livingstone Smith A (2016) Pottery and politics: making sense of pottery traditions in central Africa. Cambridge Archaeol J 26(3):471–491

Mayer E (1971) Un carnero por un saco de papas: aspectos del trueque en la zona de Chaupiwaranga, Pasco. Revista del Museo Nacional, XXXVII, pp 184–196

Mayer E (1985) Production zones. In: Masuda S, Shimada I, Morris C (eds) Andean ecology and civilization. University of Tokyo Press, Tokyo, pp 45–84

Mayer E (2002) The articulated peasant: household economies in the Andes. Westview Press, Boulder

Mayer E, Alberti G (eds) (1974) Reciprocidad e intercambio en los Andes Peruanos. Instituto de Estudios Peruanos, Lima

McCarthy J (2006) Neoliberalism and the politics of alternatives: community forestry in British Columbia and the United States. Ann Assoc Am Geogr 96(1):84–104

Mohr KL (1992) The organization of production and distribution of traditional pottery in South Highland Peru. In: Bey GJ III, Pool CA (eds) Ceramic production and distribution. Westview Press, Boulder, pp 49–92

Monzón S (1991) El estudio de la cerámica y su contribución a una investigación interregional - el caso de Piura. BIFEA 20:589–597

Murra JV (2002 [1972]) El control vertical de un máximo de pisos ecológicos en la economía de las sociedades andinas. In: Murra JV (ed) El Mundo Andino. PUCP Fondo Editorial/IEP, Lima, pp 85–125

Orlove BS (1974) Reciprocidad, desigualdad y dominación. In: Alberti G, Mayer E (eds) Reciprocidad e intercambio en los Andes Peruanos. IEP, Lima, pp 290–321

Orlove BS (1998) Down to earth: race and substance in the Andes. Bull Latin Am Res 17(2):207–222

Perreault T (2009) Environment and development. In: Castree N, Demeritt D, Liverman D, Rhoads B (eds) A companion to environmental geography. Wiley, Malden, pp 442–460

Polanyi K (1975) Traders and trade. In: Sabloff JA, Lamberg-Karlovsky CC (eds) Ancient cvilization and trade. University of New Mexico Press, Albuquerque, pp 133–154

Ramón G (2008) Producción alfarera en Piura (Perú): estilos técnicos y diacronía. BIFEA 37(3):477–509

Ramón G (2011) The swallow potters: seasonally migratory styles in the Andes. In: Scarcella S (ed) Archaeological ceramics: a review of current research. Oxford, British Archaeological Reports, pp 160–175

Ramón G (2013a) Los alfareros golondrinos: productores itinerantes en los Andes. Instituto Francés de Estudios Andinos, Lima

Ramón G (2013b) Las fuentes del estilo: distribución regional de canteras y técnicas alfareras en Conchucos (Ancash, Perú). BIFEA 42(1):29–90

Ramón G, Bell MG (2013) Re-placing plainware: production and distribution of domestic pottery, and the narration of the pre-colonial past in the Peruvian Andes. J Anthropol Archaeol 32:595–613

Reardon T, Berdegué J, Escobar G (2001) Rural nonfarm employment and incomes in Latin America: overview and policy implications. World Dev 29(3):395–409

Reardon T, Delgado C, Matlon P (1992) Determinants and effects of income diversification amongst farm households in Burkina Faso. J Dev Stud 28:265–296

Reardon T, Vosti SA (1995) Links between rural poverty and the environment in developing countries: asset categories and investment poverty. World Dev 23:1495–1506

Rigg J (2006) Land, farming, livelihoods and poverty: rethinking the links in the rural South. World Dev 34(1):180–202

Sabogal JR (1982) La cerámica de Piura. Tomo 1. IADAP Sede Central, Quito

Salomon F (1985) The dynamic potential of the complementarity concept. In: Masuda S, Shimada I, Morris C (eds) Andean ecology and civilization. University of Tokyo Press, Tokyo, pp 511–531

Salomon F (1986) Native lords of Quito in the age of the Incas. Cambridge University Press, Cambridge

Sillar B (2000) Shaping culture: making pots and constructing households. An ethnoarchaeological study of pottery production, trade and use in the Andes, vol 83. British Archaeological Reports, Oxford

Smith ML (2005) Networks, territories, and the cartography of ancient states. Ann Assoc Am Geogr 95(4):832–849

Spradley JP (1979) The ethnographic interview. Holt, Rinehart & Winston, New York

Stern SJ (1995) The variety and ambiguity of native Andean intervention in European colonial markets. In: Larson B, Harris O, Tandeter E (eds) Ethnicity, markets, and migration in the Andes. Duke University Press, Durham, pp 73–100

Swinton SM, Quiroz R (2003) Is poverty to blame for soil, pasture and forest degradation in Peru's Altiplano? World Dev 31(11):1903–1919

Tomoeda H (1985) The llama is my chacra: Metaphor of Andean pastoralists. In: Masuda S, Shimada I, Morris C (eds) Andean ecology and civilization. University of Tokyo Press, Tokyo, pp 277–300

Tosi JA (1960) Zonas de vida natural en el Peru. Instituto Interamericano de Ciencias Agricolas, Lima

Turner S (2007) Trading old textiles: the selective diversification of highland livelihoods in Northern Vietnam. Hum Org 66(4):389–404

Turner MD, Williams TO (2002) Livestock market dynamics and local vulnerabilities in the Sahel. World Dev 30(4):683–705

Valdivia C, Dunn ED, Jetté C (1996) Diversification as a risk management strategy in an Andean agropastoral community. Am J Agric Econ 78(5):1329–1334

Weismantel MJ (1988) Food, gender and poverty in the Ecuadorian Andes. University of Pennsylvania Press, Philadelphia

Weismantel MJ (2001) Cholas and pishtacos: stories of race and sex in the Andes. University of Chicago Press, Chicago

Wolf E (1966) Peasants. Prentice-Hall Inc.,Englewood Cliffs

Zimmerer KS (2014) Conserving agrobiodiversity amid global change, migration, and nontraditional livelihood networks: the dynamic uses of cultural landscape knowledge. Ecol Soc 19(2):1

Chapter 2
Networks of Pottery Exchange

Abstract This chapter describes the pottery distribution methods used in San Bartolomé de los Olleros. The most important ways that pots are distributed are: (1) trading trips to other communities to barter pots for the products of those communities; (2) sale/barter of pots to customers who visit potters' homes, including: customers within the community (community-based exchange), customers purchasing for their own use (ad hoc trade) and bulk purchases by middlemen and women who take the pots to other communities to redistribute. The methods employed less frequently are annual fairs, itinerant production, and markets (weekly and daily). Maps of each trade method are presented and specific trade routes are explained.

Keywords Barter · Trade · Pottery · Andes

2.1 Methods of Pottery Distribution and the Case of San Bartolomé de los Olleros

Polanyi's classic prescription for understanding trade serves as a basic foundation for classifying the pottery exchange methods used by potters from San Bartolomé de los Olleros. He writes:

> Since something must be carried by someone over a distance and this is to happen in two directions, trade must involve (1) personnel; (2) goods; (3) carrying; and (4) two-sidedness. All of these institutional features permit classification according to criteria which are either sociological or technological or both (Polanyi 1975: 136).

To trace Olleros pottery trade networks, and thus to reconstruct the spatiality of pottery-based livelihoods, the major categories of people involved, products exchanged, transport methods used, and social interactions are all considered.

Thus far, few scholars of Andean pottery have paid close attention to the distribution of the final product. Among those that do, most notably Mohr (1992) and Sillar (2000), the focus has been given to examples from Southern Peru (Cuzco and Puno) and Bolivia (Cochabamba and Potosi). However, Sillar provides a useful

M.G. Bell, *Pottery, Livelihoods, and Landscapes*, SpringerBriefs in Latin American Studies, DOI 10.1007/978-3-319-52331-6_2

Table 2.1 Pottery distribution methods in Peru, taken from Sillar (2000: 87)

	Pottery trade method
0	Production for own use: pots are not exchanged beyond the potter's household
1	Community-based exchanges: pots are exchanged within the potter's community
2	Ad hoc trade from the potter's house: consumers come directly to the potter's house to acquire pots for their own use
3	Weekly markets: potters and/or middlemen or women trade pots in the village square
4	Annual fairs: festivals where many potters and/or middlemen or women trade pots
5	Itinerant potters: potters make vessels in consumers' communities and return home with traded goods
6	Traveling traders: potters and/or middlemen or women take fired pots to other communities to trade
7	Bulk purchases: Middlemen or women buy pottery, occasionally made to order from potters
8	Daily markets: largely middlemen or women selling a range of wares from a variety of production centers

framework for examining pottery distribution in other Andean regions; in particular, he identifies eight distinct ways pots may be traded. He distinguishes these by three factors: (1) who is trading, (2) where trading takes place, and (3) the timing of the trade (daily, weekly or annual). This study draws on Sillar's (2000: 87) concepts and terminology, which are summarized in Table 2.1.

All of these trading methods can be observed in Northern Peru. There, just as both Sillar (2000) and Mohr (1992) observed for Southern Peru and Bolivia, the importance of these methods varies between pottery production centers as well as between potters within a single community. So, for example, traveling trade (6) is the dominant method in some communities, whereas annual fairs (4) are the most important trading venues in others.

In Olleros, not all of the pottery exchange methods described in Sillar (2000) are used. Table 2.2 lists the methods that are used, classifying them by their varying levels of importance. While these generalized categories are helpful for

Table 2.2 Pottery distribution methods in Olleros, methods within each column are listed in the order that they are discussed below

Very important	Occasionally practiced	Never (or rarely) practiced[a]
Production for own use (0)	Annual fairs (4)	Weekly markets (3)
Traveling traders (6)	Itinerant potters (6)	Daily markets (8)
Community-based exchange (1)		
Ad hoc trade from the potter's house (2)		
Bulk purchase (7)		

[a]Small local markets (weekly or daily) do not exist in Olleros or any of the communities immediately surrounding it, probably because of the proximity to the major city of Ayabaca (about 5 h walking distance). Pots are sometimes sold in the main Ayabaca market in market women's stalls, this type of exchange is described in the text

demonstrating basic trends within a community, more detailed description is needed to understand how each method is actually used.

To begin, while potters do use their pots for cooking in their own homes, their major reason for making pots is to trade them for goods they cannot or do not produce for themselves, including money. The most important ways that pots are distributed are: (1) trading trips to other communities to barter pots for the products of those communities; (2) sale/barter of pots to customers who visit potters' homes, including: customers within the community (community-based exchange), customers purchasing for their own use (ad hoc trade), and bulk purchases by middlemen and women who take the pots to other communities to redistribute. The methods employed less frequently are annual fairs, itinerant production, and markets (weekly and daily). In the following paragraphs each of these methods is thoroughly described, and the spatial patterns of distribution resulting from each method are discussed. These descriptions are meant to be of typical examples of each method; they were created by combining many traders' responses.

2.2 Trading Trips

Almost all potters do, or have, traveled to other places to barter or sell their pots. A trading trip usually begins with a potter and/or her husband, (or father if she is young), deciding on the place(s) to travel to, and then finding traveling companions, usually a family member or friend from the community. Women potters do not rely on male family members to trade, although they will not travel long distances alone. Pottery traders travel to several different types of places, but generally they seek places where grain production is abundant (termed *sitios graneros*, see Sect. 3.1) and they especially desire to barter their pots for wheat, maize, peas, and beans. Trade destinations are usually specific *caseríos* (sectors) of *campesina* communities, the clustering of houses in these settlements makes trading easier (see Appendix C for a registry of trade destinations).

Trading trips only occur between the months of July and November, with a peak in August. This occurs for two reasons. First, because travel is much easier during the dry months. In fact, some of the trade routes between communities in the highlands of Piura are almost completely cut off during the rainy season, causing communities to stock up on trade goods during the dry season (Córdova, pers. comm.). And second, traders follow the harvests in the locations where they trade. During and directly after this time the exchange rates are best (see Sect. 3.2). Harvest timing varies between farming communities. According to Mauro Mondragón, an experienced trader, the first harvest among the distribution sites begins in July and the last ends in September. Once the destination and date of the trading trip have been determined, the traveling party collects donkeys to carry the loads. These donkeys may be the trader's own, or they may be rented or borrowed. The load each donkey can carry is called a *carga*, twelve pots are in one *carga* of pots. Usually one to five donkeys per pottery trader are used, however not all of the donkeys are loaded with pots. Other products, especially *chancaca*, are often brought for trade as well. Traders travel up to two days walking (at a donkey's pace) one way. They sleep at fixed points en route, which are determined based on the availability of pasture for the pack animals.

Most pottery traders give similar accounts of their arrival in destination communities. When they arrive they call out "pots, pots" ("*ollas, ollas*") and people come outside to trade. Pots are usually bartered for potfuls of grain or other products, although sometimes they are sold for money (see Sects. 3.1 and 3.2). Almost all traders have friends or acquaintances (*amistades, conocidos*) in the places where they trade, these are partnerships that develop over years of visits. Potters greatly appreciate trading with these established friends, but this by no means limits exchange with new, as yet unknown, partners. For more distant destinations, traders will spend the night in the homes of these friends. For closer destinations, departure, trade, and return are all accomplished within a single day. Traders generally continue traveling from village to village until they trade all of their pots; no one wants to return home still carrying the load with which they set out. Sometimes, however, traders will miscalculate harvest times or deliberately arrive out of harvest. In these cases, pots are left with trade partners who promise to return the equivalent quantity of grain when they have it. This speaks to the level of trust inherent to these exchange relationships. Additionally, while the pots that are traded in destination communities are often for use in the trading partner's household, sometimes individual households will trade for up to six or eight pots and then re-trade these pots in still more distant places.

The most important site traveled to from Olleros is Matalacas, an immense grain-producing community about a two days' walk away. However, Matalacas is only one destination among many. It is perhaps most useful to classify destinations according to routes traveled, including all the stops made along the way, than to just list destinations as individual entities. In fact, almost all trading trips follow one of three distinct routes, each of which is fully described below, and represented in Fig. 2.1.

Route 1: Southeast to the community of Matalacas. (*34 references to different sectors in interview data*)

Matalacas is by far the most important exchange center for Olleros potters, mainly because it is a highly fertile grain producing zone, possessing far better and much more vast agricultural lands than do other parts of the Ayabaca highlands. According to Hildegardo Córdova, a prominent Peruvian geographer and native of the Piuran sierra, Matalacas is a regional bread basket (pers. comm.). It yields some of the highest volumes of wheat in Piura; cultivating at least thirteen different wheat varieties, mostly without external additions like pesticides (Bernex 1990: 98), and without irrigation ("*solo secano*" Córdova, pers. comm.).[1] Generally, wheat is

[1]At least this was the status of agriculture in the late 1980s, changes since then would not be surprising.

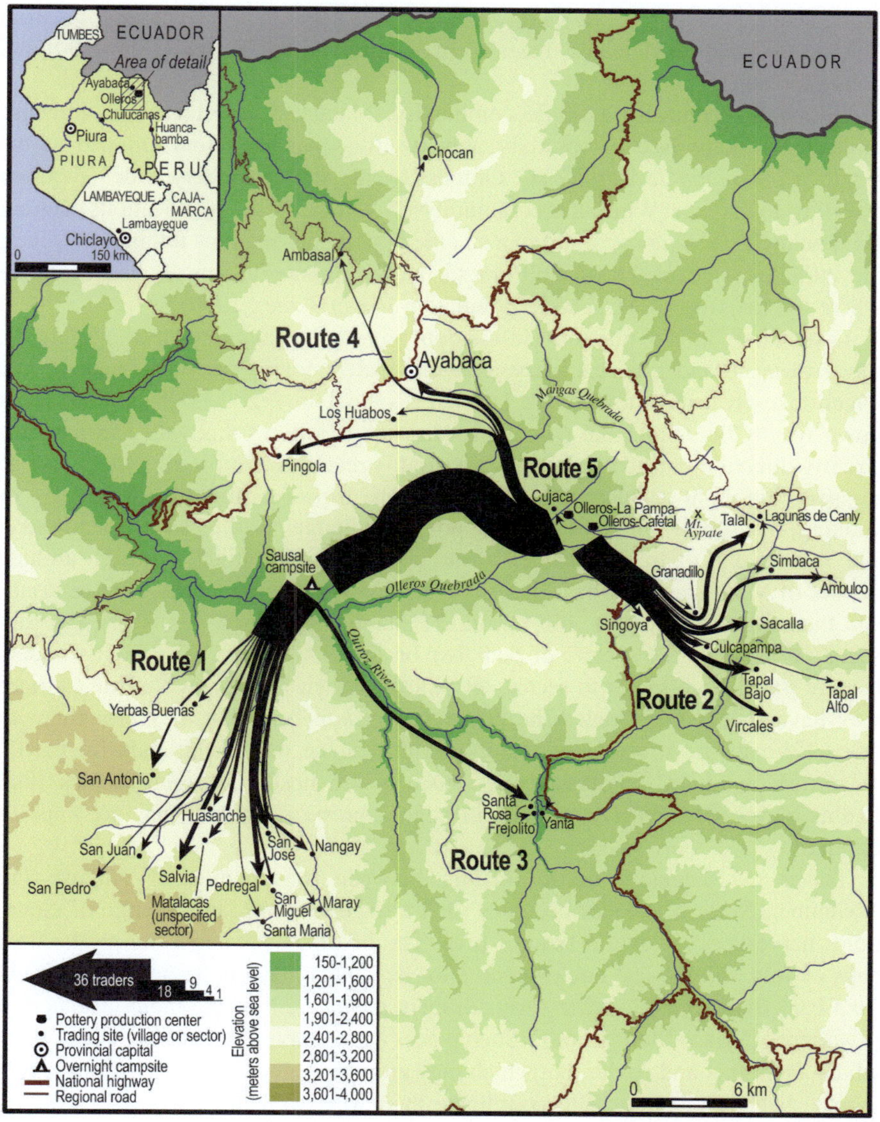

Fig. 2.1 Trading trip routes and destinations. Flow lines represent approximate routes followed, line thickness represents the number of traders who mentioned traveling to a specific destination during interviews

cultivated across a range of elevations, from 1600 to 3000 m.a.s.l. It is mixed with potatoes and barley at the higher end of this range and with maize at the lower end (Bernex 1990: 98–8). Pottery traders from Olleros concur with this description of Matalacas, as potter Emilia Pintado put it: “a lot of people go [to Matalacas] to exchange for grain; peas, wheat, potatoes, maize (*maíz amarillo*), beans, wool, broad beans, everything! The land is different, it is better there."[2] Trading pots from Olleros for grain in Matalacas has also been observed by Bernex (1990: 100) and Sabogal (1982: 129).

Although technically a single community, Matalacas is actually an immense ex-hacienda comprised of approximately twenty separate sectors grouped as the single "*Comunidad de Matalacas.*" Of these twenty sectors, pottery traders reported traveling to sixteen. While the community of Matalacas spans elevations ranging from below 1600 to above 3500 m.a.s.l., the potters' destinations fall mainly within the altitude range of 2400–3200 m.a.s.l. Not surprisingly, this is also the zone where the majority of Matalacas' population lives (Bernex 1990).

Depending on the ultimate destination within the community of Matalacas, travel distance and time vary. Even so, all traders follow the same major route which takes two days to traverse (refer to Route 1 in Fig. 2.1). This route leads down from Olleros into the *Quebrada Mangas*. It follows this valley in a south-easterly direction until the campsite at Sausal is reached. Almost all traders choose to spend the night at this point. This leg of the trip takes less than a full day of walking, but Sausal is a good stopping point as suitable pasture can be found there. The second day of travel includes a long ascent into the middle to high altitude sectors of Matalacas: "we pass that river and cross to the other side there [Sausal], and then we make it up the hill" (Emilia Pintado).[3] Common points of entry into Matalacas include the sectors of San José or San Juan. From these points traders usually travel from sector to sector, all following their own particular routes, until they have successfully traded the pots (and/or *chancaca*) that they brought with them. "We go from sector to sector like this, since one has friends [in each]" (Mauro Mondragón).[4] Pottery traders usually are hosted by their friends (and trading partners), they may stay several nights, presumably until they are finished trading.

Travel to Matalacas occurs mainly between the months of August and November. Since Matalacas occupies such a huge territory, and a range of elevations, harvests occur over a span of time. The first harvests begin during the final days of July and the last one extends into September. This is because lower parts (like Palo Blanco) are warmer: "we call them *temple*, …they are more insulated, it is warmer" (Mauro Mondragón).[5] Such places begin harvesting earlier, while others which are higher and colder, but which may grow the same crops, harvest up to two months later. Traders have to coordinate their trips with this geography of agriculture. During the harvest Matalacans are willing to barter their grain for pots and exchange rates are good. Afterwards it seems that they prefer to sell their crops for money, a situation not very beneficial to potters (see longer discussion in Sect. 3.2).

[2]"…van bastante gente a cambiar con grano: alverja, trigo, papas, maíz amarillo, fríjol, papa, lana, habas, todo!…. La tierra es diferente, allá mejor" (Emilia Pintado).

[3]"Pasamos este río y bandeamos allá, y de ahí ganamos el cerro" (Emilia Pintando).

[4]"Vamos a sector a sector así, como uno tiene amigos" (Mauro Mondragón).

[5]"…llamamos temple…son mas abrigados, es mas caliente" (Mauro Mondragón).

It is fair to say that economically speaking, Matalacas is more important to Olleros than Olleros is to Matalacas (with the possible exception of *chancaca* production). This asymmetrical relationship directly affects trade patterns, so a basic understanding of some of the historical and economic trends influencing this relationship is relevant.[6] The Hacienda Matalacas, which was in existence as early as 1615 (Schulpmann 1994, Annex 2: 24), was devoted to raising livestock (Bernex 1990: 95). Following the lead of several other estates in the region, in 1953 the owners of Matalacas decided to sell the land to the hacienda workers, who together purchased private plots in a single collective movement (Apel 1996: 61). This occurred nearly twenty years before the national Agrarian Reform Law in Peru, after which the Hacienda Olleros was broken up.[7] This created a large group of smallholding farmers, many of whom began to cultivate the wheat Matalacas is known for today. Thus the change in the land tenure system also resulted in a significant change in regional land use. Since the 1950s, through a combination of agriculture, herding, and commerce, Matalacans have become relatively wealthy; consequently, they are a political power in the region. In fact, they represent strong enough of a political force to influence road construction routes. As early as the 1960s, Matalacas was directly linked to the coast by a road passing through the sectors of Las Pircas and Salvia (Franco 1990). It is via this route that much of the grain from Matalacas reaches the coastal markets.

While this coastal orientation and market integration is recognized by the pottery traders, they have mixed responses regarding its influence on their activities. For most it does not seem to have had a significant impact on trading practices. Even the sector of Salvia, where "the roads enter" (José Adán Troncos),[8] is still one of the most popular destinations for pottery traders from Olleros. It is also worth noting that Salvia is one of the more highly populated centers within Matalacas. Yet, several traders have commented on the changes that the Salvia/Las Pircas road and consequent market expansion have wrought on local systems of barter exchange, perhaps referring to slowly shifting patterns. "Now things have changed because there are roads which allow [products] to be taken to the [coastal] cities. So now the people [from Olleros] have to go even farther, where the roads still don't reach"

[6]Most of the following information comes from a large study of the central highlands of Piura carried out in the 1980s by the Pontifical Catholic University of Peru and ORSTOM (Institut français de recherche scientifique pour le développement en coopération), and led by N. Bernex, H. Cordova, J.C. Roux, and G. Etesse. The detailed discussion of Matalacas presented here would not be possible without this study. Since such detail does not exist in the literature about the other trade destinations, an imbalance in this text is unavoidable.

[7]"The Agrarian Reform of 1969 manifested itself slowly and reluctantly in Piura. In 1970 it had only affected some of the largest estates on the coast located in the Chira valley and until 1974 almost the entirety of the Department's sierra region was excluded from its actions". (Rubin de Celis 1978: 11 cited in Apel 1996: 78, my translation).

[8]"…entra allá carretera de carro…" (José Adán Troncos).

(Beltrán Mondragón, son of Mauro Mondragón. Mauro is a prominent trader, who interestingly has no such complaints about Matalacans).[9]

Some anthropologists similarly argue that in Matalacas the traditional *campesino* subsistence economy is being abandoned for a more market oriented approach (Franco 1990: 325). This can be debated, since certainly it is "still" possible now to see the "old" barter circuits coexisting with the capitalist system (Bernex 1990: 100). For example, at least fairly recently, Matalacas was found to be strongly integrated into regional trade networks. Bernex (1990: 100) provides a vivid description of how traders from the entire region surrounding Matalacas go there with products to trade for its wheat and other crops. This includes traders from Morropón, Chalaco, and Pacaipampa who bring salt, soap, kerosene, rice, and sugar; from Aragoto carrying *bocadillos* and *alfeñiques*[10]; and of course from Olleros hauling clay pots. Matalacans themselves also travel on trading trips to other places. For example, from the sectors of Nangay and San Miguel they take tanned cow skins to the coast (Sullana). Additionally, from many of the higher altitude sectors traders go to Olleros (and other sugarcane growing zones) looking for *chancaca*, which they either cash or trade for with grain they bring with them. Interestingly, Matalacans rarely travel to Olleros looking for pots, the real attraction for them are the sugarcane products. This once again implies that pottery trade relationships are asymmetrical, with Matalacan farmers holding greater power than Olleros potters, perhaps because of Matalacas' potential for increasing commercialization and abandoning "traditional" exchange systems. However, it also demonstrates the continuity of *campesino* barter even amidst increasing market integration, indicating the continued importance of this trade to rural livelihoods.

Route 2: Passing El Aypate: East to the communities beyond Mt. Aypate (*30 references to different villages and sectors in interview data*)

A host of destinations for pottery traders are located to the east of Olleros, beyond the peaks of mounts Aypate and Viscacha, which form one boundary of the community. While the elevations of these destinations vary considerably from 2000 to 3600 m, these places are all grain producing zones (see Sect. 3.1). The main products bartered are maize, peas, beans, and wheat. Unlike the multi-destination routes followed during trips to Matalacas, these destinations are mainly visited in discrete trips. They can all be reached in a single day of walking. With trading, the total trip time is usually one to three days. Due to the proximity of these sites, (especially Tapal, which is said to be the shortest trip), these places are easier to

[9]"ahora cambió porque hay carreteras, y todo eso lo acapara y lo llevan a las ciudades [en la costa]. Entonces ahora la gente tiene que irse mas lejos donde no hay carros donde no entran las carreteras ya" (Beltrán Mondragón).

[10]*Bocadillos* are a kind of sweet made from *chancaca* and peanuts. *Alfeñiques* are a kind of molasses taffy. Both are commonly sold in festivals and marketplaces throughout Piura.

access than Matalacas. Since they grow the grains the potters seek, they are very popular destinations, especially among traders who are unable to travel the long distance to Matalacas, such as the elderly (see Sect. 3.3). The communities in this group are almost all ex-haciendas, which gained independence at varying times between the late 1940s and early 1970s (Apel 1996). Each pottery trader has specific communities among this group that he or she tends to visit; although traders do not always return to the same place every year. It is not clear why some traders visit some sites and not others; but, since traders have friends in each of the communities they visit, this may be a defining factor in their decision to continue visiting a place again and again.

Route 3: Lowlands to the south: Santa Rosa and Yanta (*4 references in interview data*)

These two sites located to the south of Olleros (Santa Rosa and Yanta) are infrequently visited, but are significant because they are destinations known for being places to trade pots for animals. One potter (Emma Mondragón) described them as "Goat Zones" (*Zonas Cabrillas*) (see Sect. 3.1). She used the word "*cabrilla*," to reference the popularity of livestock, especially goats, which can be found in this area. Santa Rosa is located a half day's walk away, with Yanta located just upstream along the Santa Rosa river. These are relatively low (<1200 m.a.s.l.), warm locations, only visited from time to time.

Route 4: Passing Ayabaca: The city of Ayabaca and surrounding communities (*11 references in interview data*)

Ayabaca is a major regional urban center. It is the provincial capital, the site of a large annual Catholic pilgrimage to "*El Poderoso Señor Cautivo de Ayabaca,*" and also home to a permanent daily market. Olleros pots are sometimes sold in the market (see Sect. 2.7). Olleros is the only potting community which distributes in this urban market. Selling or trading pots in the city is relatively rare. Potters who travel in this direction generally use Ayabaca as a waypoint; either for catching a ride on the way to other destinations (e.g., Pingola), or along a route traveled by foot, to the north (e.g., Ambasal en route to Chocán) and to the southeast (Pingola, La Laguna, Los Guavos). The products available at these higher altitude destinations include peas, maize, wheat, and broad beans. These are not common destinations, and the one potting family who described this route has not traveled to these places "in years" (Orfilia and Mauro Mondragón).[11] Overall they are more or less abandoned routes. The only trader from Olleros who has traveled once recently is not actually a potter herself: Altagracia Chuquihuanca visited Pingola for the local fiesta in 2005 (November 20), traveling by car with a half-*carga* of pots, which she traded with friends at the fiesta for maize and wheat.

[11] "hace años" (Orfilia and Mauro Mondragón).

Route 5: Within and directly surrounding the community of Olleros (*3 references in interview data*)

Several pottery traders mentioned occasional visits to the non-pottery producing sectors of the community of Olleros, specifically the sectors Congoli and Toronche, which are located about two hours walking from La Pampa (slightly less from Cafetal). Also included in this "route" are visits to the neighboring community of Cujaca, which is very near to La Pampa (approximately 30 min walking, slightly longer from Cafetal, see Figs. 1.3 and 2.1). During these trips pots are traded for special products eaten fresh, like *choclo* (fresh corn on the cob), or products falling under the general category of "*Recado*" which in Olleros is a food type referring to products such as manioc or plantains (see Sect. 3.1).

Trips to these places have an "occasional" air to them. For example, one potter, Micaela Jimenéz, traveled to Congoli during the fieldwork period at the invitation of a friend. She left at 3 AM and arrived at 6 AM. She took her son and three pots that she had "borrowed" from her neighbors since she did not have any fired pots at the time. She left the pots with her friends and in return they gave her two *alforjas* (saddle/shoulder bags) full of *choclos* as well as some beans. She also helped in the harvest and was given additional *choclos* in return (see Fig. 2.2). Micaela passed the better part of one day on her trip Congoli, and she returned in the evening to roast fresh *choclo* with her children (Fig. 2.3). Another potter, Zenaida Huaman, recounted that she occasionally travels to Congoli and Toronche with her mother-in-law (potter Rosa Jiménez) to trade for *recado*.

Fig. 2.2 Micaela Jiménez "borrowing" pots from one of her neighbors, in Olleros-La Pampa. Mt. Aypate and Olleros-Congoli are visible in the background

Fig. 2.3 Micaela Jiménez and her daughter preparing the choclo brought from Olleros-Congoli in exchange for pots, in Olleros-La Pampa

While these locations within Olleros are easy destinations, trips to them are infrequent and do not account for the majority of pottery trade. Potters cannot acquire the staple foods they need by exchanging pots within their home community, even in the higher elevation sectors—this is the main motivation for the trading trips described in Routes 1–4.

2.3 Community-Based Exchange

Potters and their family members are not the only ones who travel from Olleros to other places to trade. Pots are regularly exchanged within the community of Olleros between potters and other community members who do not practice this craft. These non-potters then (re)trade the pots in the same ways that potters themselves do. Pots are lent/borrowed (*se presta*) or sold/bought (*se compra*) between potters and other potters or between potters and non-potters. Pots are usually lent when the borrower is planning a trading trip and wants to take more pots for trade than he or she has at the moment. The borrower returns new pots to the lender when he or she is able, directly replacing those initially borrowed. When pots are bought, they can be paid for either in cash or kind, the term "buy" (*se compra*) is used to describe both forms of payment. The products used as payments in such cases can vary, and seem to be flexibly determined by the potter and buyer. Examples include textiles (Altagracia Chuquihuanca trading *alforjas* for pots from Micaela Jiménez), *chancaca* (Orfilia Mondragón), and wool (Irma Abad).

These trades often occur between *comadres* or close friends. All the examples of intra-community exchanges collected during fieldwork interviews were between women, but presumably Olleros men could also buy pots from potters. The general consensus among potters and non-potters is that "… some know how to [make pots], and others buy" (Oralia Mondragón) or "When someone doesn't know how, she buys from the women who do know" (Honoria García).[12] Pottery buyers may also store pots in their homes for future trading trips or to sell to visiting customers (see Sect. 2.4).

2.4 Visiting Customers: Ad Hoc Trade from the Potter's House and Bulk Purchase

Customers from beyond Olleros also come directly to the potters' houses for pots. These customers range from individuals looking for one or two pots for their own kitchens, to bulk purchasers who buy up to fifteen *cargas* (180 pots) from several potters at once, which they then redistribute. Many of the bulk traders come from the communities directly surrounding the city of Ayabaca. However, customers hail from many different locations, including many of the places where potters themselves travel to trade (see Appendix C for a registry of customer points of origin).

The potter Luzmila Parihuamán describes a typical customer visit as follows. The customer arrives to her house, she usually knows him (or her), and they greet one another outside. The customer(s) asks if she has pots, she replies yes and asks them what he or she has brought in exchange. The potter accepts (or presumably could refuse) the transaction and they exchange with one another.

Customers may bring both goods and money to exchange for pots. Potters say that customers bring "any kind of thing" for trade. This includes lowland foods (manioc, bananas, sugar, cacao), grains (wheat, maize, barley), *menestras* (beans, peas), products from the Ayabaca city market (kerosene, salt, soap, fish, vegetables), animals (guinea pigs, chickens), wool, textiles, and more. Customers can come casually while they are passing through Olleros, or they may order specific

[12]"Es que algunas saben hacer, y otras compran" (Oralia Mondragón) or "O sea cuando uno no sabe hacer compra a las señoras que saben hacer…" (Honoria García).

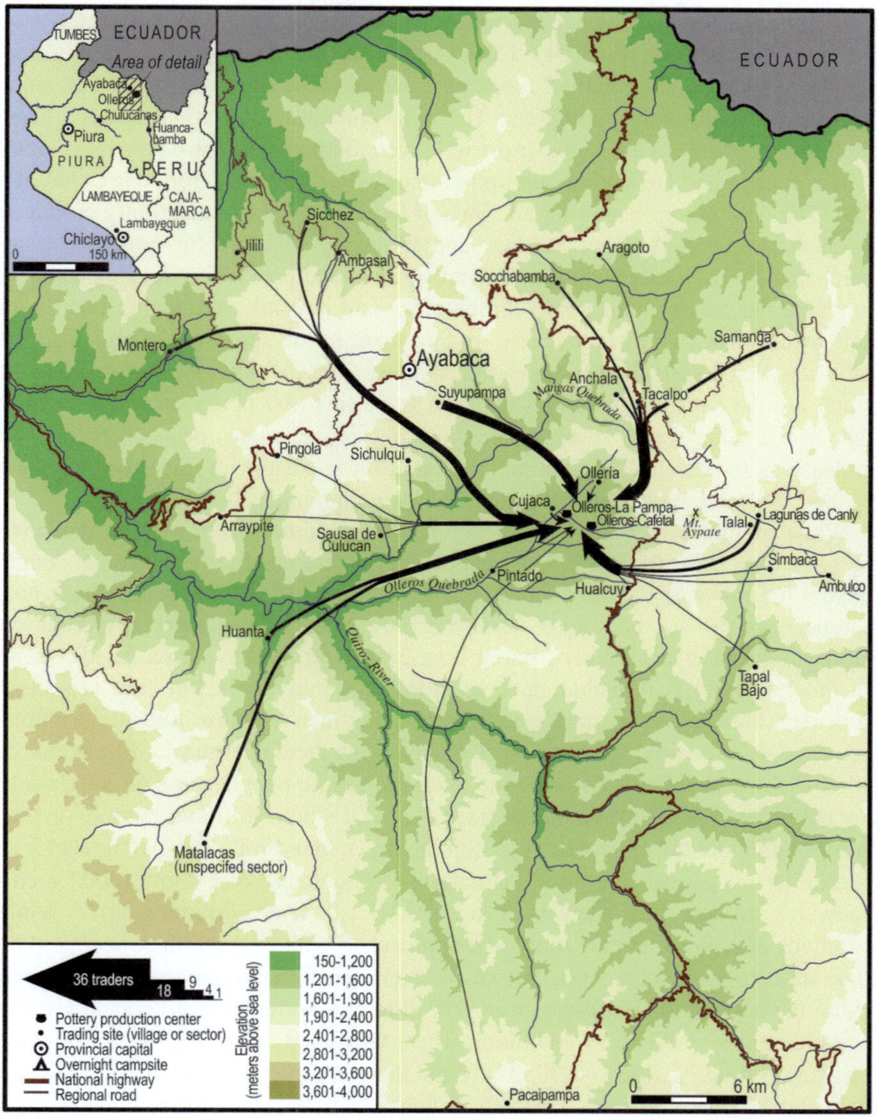

Fig. 2.4 Points of origin for visiting customers. Flow line thickness approximates the number of traders traveling from each destination (represented by the number of times a specific location was mentioned in interview data)

quantities of pots ahead of time. Customers arrive especially during the months of July and August, some plan their visits to coincide with Olleros' community festival (San Bartolomé, August 24). Their visits may also coincide with festivals in other communities if they are planning to re-sell/re-barter the pots in those places.

Visiting customers come from many of the communities in the region surrounding Olleros (see Fig. 2.4). Customers from the communities directly surrounding Ayabaca, especially Suyupampa, are arguably the most important, since they are often the bulk purchasers. The bulk purchasers are popular individuals, referred to by name. This is not surprising given that they purchase pots from many potters at once and are said to buy up to three *cargas* (36 pots) from one potter at a time. The bulk purchasers bring all kinds of foods, as well as industrially produced products from the Ayabaca market, and/or money to trade for pots. They are merchants (*negociantes*) who seek to redistribute the pots. After leaving Olleros, they take the pots to their home communities and beyond, exchanging them with their own family members, but also trading them in the surrounding region in the same way that potters themselves do. Customers from the other locations, (many of which are the same communities as those traveled to by pottery traders), usually trade at a smaller scale. These customers bring grains, beans, peas, chickens, and guinea pigs for barter. The assumption is that these customers trade for their own use and for some redistribution in their home communities.

While trading trips from Olleros to other communities seem to have remained the same over time, there is a consensus within Olleros that fewer customers come looking for pots now than in the past, and specifically the pre-Agrarian Reform past (estimates of 30–40 years ago). Customers are said to have especially come on Sundays, not in great numbers, but at least 3 or 4 per Sunday (Irma Abad). Some customers still do visit, including some of the bulk purchasers, but there is general agreement that the pottery market used to be better than it is now. Such descriptions include: "…in the time when we were making pots there in Aguayco [the sector of Olleros where people lived during the hacienda], more people came on Sundays to buy pots" (Irma Abad); "Well, now since….now it's been thirty years more or less, because after that the people stopped coming down here [to La Pampa, the sector of Olleros where people live now]" (Luzmila Parihuamán); and "now [pots] aren't made, it's because now there isn't any market…" (Oralia Mondragón).[13] While narratives of a glorified past should be approached with caution, there seems to be some truth to such statements. Before the Agrarian Reform the settlement pattern within the community of Olleros was very different. Potters refer to traders visiting their old homes, in Aguayco and Sidro, where they lived during hacienda times (see Fig. 1.3). Similarly, Roux (1990: 32) found that caravaneers in the central sierra of Piura thought haciendas to be good trading places because many workers lived close together and were interested in small scale trade. They feared the decline of their profession with the breakup of the estates.

[13]Opinions include: "…en tiempo que hacíamos ollas allá en Aguayco [sector of Olleros where people lived during the hacienda], el día domingo venía más gente a comprar ollas." (Irma Abad); "bueno, ya hace…ya como treinta años más o menos porque de allí esa gente ya dejo de venir acá abajo…" (Luzmila Parihuamán); and "ya ahorita ya no se hace, es por el motivo que ya no hay negocio…." (Oralia Mondragón).

2.5 Annual Fairs

Annual fairs play a minimal role in Olleros pottery distribution. Sometimes potters plan their production to coincide with annual fairs in either Olleros (San Bartolomé, August 24) or in other communities within the distribution area, (Pingola, Sicches, and Montero), but not nearly to the extent described by Mohr (1992) (see Fig. 2.5

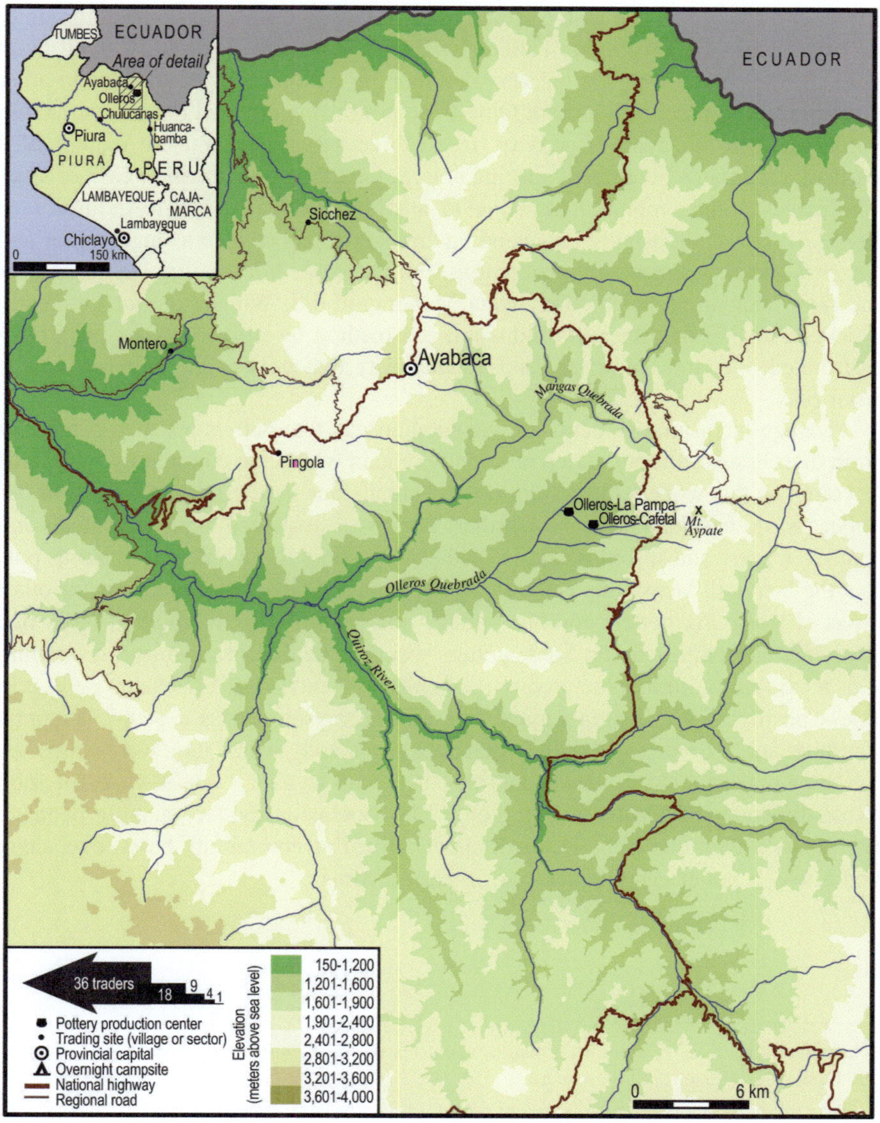

Fig. 2.5 Locations of annual fairs important for pottery trade

for locations of these fairs). Trade during festivals in Olleros occurs in the same way as described above for visiting customers, and in other communities in the same way as normal trading trips.

2.6 Itinerant Potters and Peonage

Itinerant potters are those who travel to locations away from their home community to make pots. These potters, called "swallows," are described fully by Ramón (2011, 2013). In Olleros it is not very common for a potter to have traveled to other communities to practice her craft. Only three potters interviewed mentioned having participated in this type of activity. In all three cases it was at the request of a particular woman who ordered specific pots, paying the potter in either cash or kind. The places potters reported traveling to are Pintado, Cujaca, and Andurco (see Fig. 2.6). These are usually longstanding arrangements between a single potter and her customer, even spanning generations. For example: Micaela Jiménez works for the same woman as her mother Juana Mondragón did. While it is not frequently practiced in Olleros, itinerant production is a method by which pots are physically distributed over the landscape, and another way that potters can earn an income from pots.

Peonage labor arrangements are yet another way that potters produce pots away from their homes, although they physically remain within the community of Olleros to complete this work. This system usually entails one potter inviting several of her colleagues to her home to work together. The host is expected to provide meals and a day's wages, and in return the invited potter works the whole day making pots that the host will keep. This is the same as peonage or day labor practices for other types of work. Six potters and one potter's son (referring to his mother) described participating in this activity. Some potters seem to like this arrangement, describing it as a case of traditional reciprocity ("labor/energy exchange").[14] Others paint it in a less positive light "I didn't much like renting myself out as a peon" (Luzmila Parihuamán).[15] Some say that they practiced this activity when they were single but not after marriage (Irma Abad), and that it is way to gain a day's wages or food (Rosa Jiménez and Orfilia Mondragón).[16] In conclusion, pottery-making peonage represents a way that pots are distributed, a form of community-based exchange, and a way a potter can use her craft to earn income.

[14]"cambiaron fuerzas" (Daniel Mondragón, about his mother), "nos hemos cambiado de fuerzas que llamamos" (Luzmila Parihuamán).

[15]"Poco me ha gustado así, alquilarme de peona" (Luzmila Parihuamán).

[16]"…por ganar cinco solcitos y ganarme la comidita" (Rosa Jiménez) and "Me pagaban cositas que uno no tenía pues" (Oralia Mondragón).

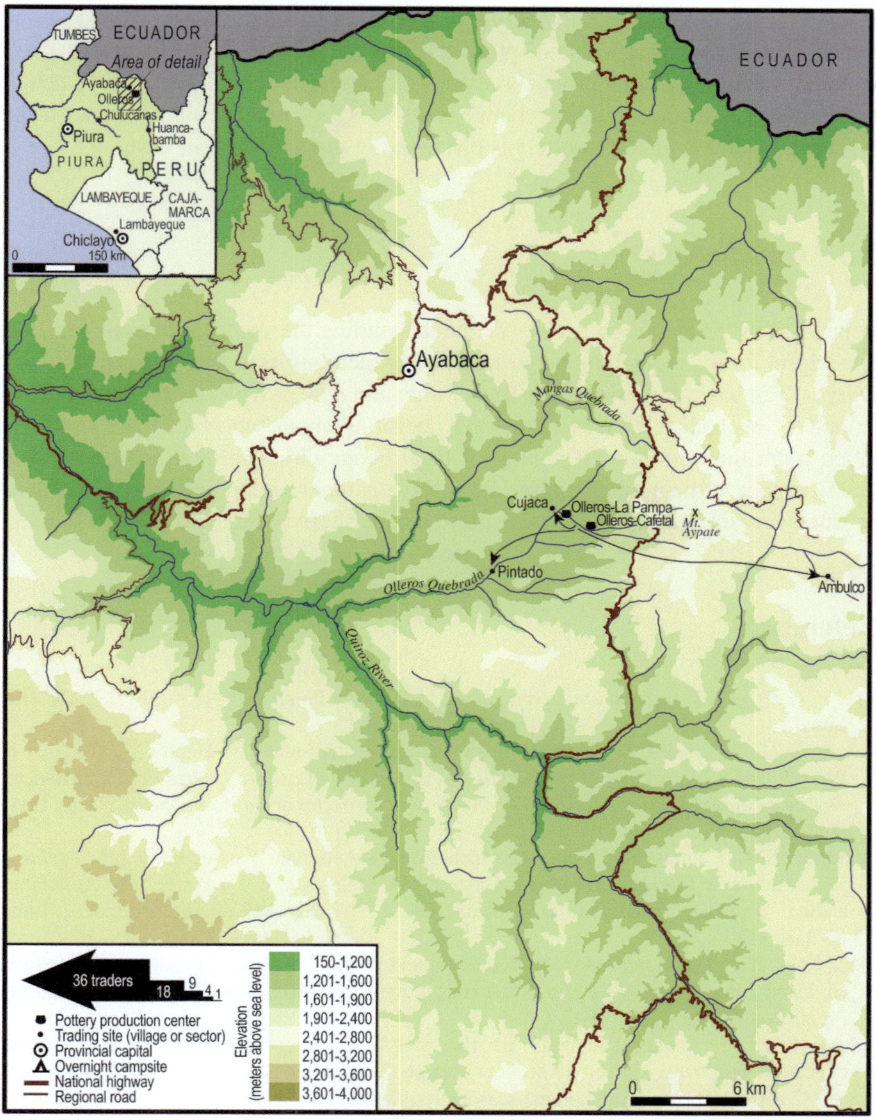

Fig. 2.6 Locations of itinerant pottery production

2.7 Urban Markets

Nowadays, sale in the urban market of Ayabaca is rare, but there are a few market stalls which sell pots from time to time. This is the result of trade between one specific potter (Micaela Jiménez) and the market women. When Micaela brings her pots to Ayabaca, she takes them directly to the market stall to sell. The vendors sum up the price of the pots. While they sometimes pay her in cash, often Micaela will take home the equivalent monetary value of products from the market stall. Micaela probably would have bought similar products anyway (e.g., rice, noodles, soap), but the benefit to the market woman is in this exchange is obvious. The relative unimportance of the Ayabaca market to Olleros potters is discussed in greater detail in Chap. 4.

2.8 Conclusions

Polanyi's (1975) classic categories for understanding the specifics of trade practices are helpful for distinguishing between different methods used to exchange pots from Olleros. The basic analysis of *who* trades *what* and *how* provides a good introduction to pottery exchange. However, making decisions about where to trade implies a much more thorough understanding of local geography, ecology, and agricultural practices, not to mention an appreciation for consumer preferences in terms of both products and exchange rates. These more detailed topics are analyzed in Chap. 3.

References

Apel K (1996) De la hacienda a la comunidad: la sierra de Piura 1934–1990. IEP/IFEA/CNRS, Lima

Bernex N (1990) El distrito de Pacaipampa: unidad y diversidad. In: Bernex N, Córdova H, Roux JC (eds) *El proyecto de desarrollo rural integral de la sierra central del Departamento de Piura (Convenio PUC-ORSTOM) Tercera parte-Aspectos socio-económicos*, PUCP, Lima, pp 88–103

Franco E (1990) Cambios microregionales cambios en los sistemas y unidades de producción: la sierra central de Piura. In: Eresue M, Gastellu JM, Malpartida E, Poupin H (eds) Agricultura andina: unidad y sistema de producción. ORSTOM/UNALM/Editorial Horizonte, Lima, pp 318–329

Mohr KL (1992) The organization of production and distribution of traditional pottery in South Highland Peru. In: Bey GJ III, Pool CA (eds) Ceramic production and distribution. Westview Press, Boulder, pp 49–92

Polanyi K (1975) Traders and trade. In: Sabloff JA, Lamberg-Karlovsky CC (eds) Ancient cvilization and trade. University of New Mexico Press, Albuquerque, pp 133–154

Ramón G (2011) The swallow potters: seasonally migratory styles in the Andes. In: Scarcella S (ed) Archaeological ceramics: A review of current research. Oxford, British Archaeological Reports, pp 160–175

Ramón G (2013) Los alfareros golondrinos: productores itinerantes en los Andes. Instituto Francés de Estudios Andinos, Lima

Rubin de Celis, E (1978), ¿Que piensa el campesino de la Reforma Agraria? Caso Piura. CIPCA, Piura

Sabogal JR (1982) La cerámica de Piura. Tomo 1. IADAP Sede Central, Quito

Shülpmann J (1994) La structure agraire et le développement d'une société régionale au nord du Pérou. Piura, 1588–1854. Doctoral Thesis, Université Paris VII, Paris, France

Sillar B (2000) Shaping culture: making pots and constructing households. An ethnoarchaeological study of pottery production, trade and use in the Andes. British Archaeological Reports, Oxford

Chapter 3
Negotiating the Pottery Exchange Landscape

Abstract This chapter interprets the pottery distribution methods and trade destinations presented in Chap. 2. It explains the decisions Olleros potters and pottery traders make regarding the marketing of their product in terms of goods sought in exchange, ecological and agricultural patterns, regional cuisine preferences, exchange rates, measurement systems, life cycles, household capabilities, and individual and community social relations. It concludes with observations of the unique spatiality of pottery exchange as a livelihood activity.

Keywords Barter · Measurement system · Ecological zone · Crop zone · Cuisine · Pottery · Andes

3.1 Kinds of Places, Kinds of Food: The Geography of Agriculture in Piura

3.1.1 Trading Pots for Food

To begin to make sense of the observed Olleros trading patterns, we must first consider the products for which pots are traded. To this end, this section discusses the basic geography of agricultural production in Piura, and especially in the highland region surrounding Olleros. Its main focus is the food products for which pots are bartered, and where and when these products are available. It also deals with pottery traders' perceptions of their environment, and connects patterns of pottery exchange with broader discussions of Andean ecological complementarity. As in the previous chapter, the discussion here is the result of combining multiple interview responses.

Olleros pots are traded for a variety of farm products, as well as for some manufactured products and also for cash. These products can be divided into eight groups: Grains, *Menestras*, *Recado*, Animals, Animal Products, Urban Products, Other Products, and Money (see Table 3.1, also see Appendix D for a full registry of products, places, and exchange rates). *Menestras*, roughly translatable as pulses (Gade 1975: 71), is a category of food and prepared dishes that consists of different

M.G. Bell, *Pottery, Livelihoods, and Landscapes*, SpringerBriefs in Latin American Studies, DOI 10.1007/978-3-319-52331-6_3

Table 3.1 Types of products traded for Olleros pots and their relative importance (by number of references in interview transcripts)

Grains (47 references)
Grains (unspecified) (6)
Maize (18)
Wheat (14)
Highland tubers (7)
Barley (2)
Menestras (41 references)
Menestras (unspecified) (1)
Field Peas (25)
Beans (12)
Broad Beans (3)
Animals (17 references)
Poultry (7)
Goats (7)
Pigs (2)
Guinea Pigs (1)
Animal Products (10 references)
Cheese (5)
Wool (5)
Recado (7 references)
Recado (unspecified) (1)
Manioc (3)
Plaintains (2)
Sweet Potato (1)
Urban products (6 references)
Other products (6 references)
Money (3 references)

Pottery traders classify their trade destinations by the products which are available for barter

vegetables, cereals, and especially legumes. In Piura, it includes field peas, beans of various varieties, *sango de trigo* (a toasted wheat dish), *repe de zapallo* (a squash and bean dish), and more combinations (Hocquenghem and Monzón 1995: 42). *Recado* is likewise a broad grouping of foods, consisting of (generally) dry starchy ingredients (*alimentos secos*) also sometimes called *trama* or *compaña*.[1] In Piura, *Recado* includes potatoes, boiled sweet potatoes, manioc, ocas, roasted/boiled corn

[1]"Recado" is an interesting word with many definitions, several of which have culinary connotations. According to the *Diccionario de la Real Academia Española* (2004), s.v. "recado," the word comes from the verb "recadar" and means the following: (1) Message or response given or sent to someone, (2) Errand or commission, (3) Memory or remembrance of esteem or affection held about someone, (4) **Daily provision of household supplies brought from the market**, (5) Combination/grouping of objects necessary to do certain things, (6) Document that explains an account, (7) Precaution, security, (8) Group of types, signs, etc., from one document made use of in another, (9) *Arg., Bol. & Ur.* Riding gear, (10) *El Salv., Gaut., Hond. & Nic.* **Liquid seasoning or spice used as condiments for meat**, (11) *Hond. & Nic.* **Chopped meat used to fill empanadas**, (12) *P. Rico.* **Aromatic plant used as seasoning**, (13) Gift, present (my translation and emphasis).

on the cob (*choclo*), plantains (*seda* and *para freir*), toasted corn (*cancha*), and potato/sweet potato/plantain chips (*chifles*) (ibid.).

Grains and *Menestras* are by far the most common food groups exchanged for pots. Within these groups field peas, maize, wheat, and beans are the most popular products (in order of number of references in the interview data). Least popular are *recados*, urban products (e.g., kerosene, salt, and more), other products (e.g., cacao, *chancaca*, and more), and money.

Traveling traders from Olleros talk about three different types of sites or zones, these are: Grain sites (*zonas graneras*), *Recado* sites, and Animal sites (*zonas cabrillas, zonas para animales*) (see Fig. 2.1 for locations of site types and Appendix C for a classification of trade destinations by route and by site type). The "Grain" classification includes sites where pots are traded for both grains and *menestras*. These are located at higher elevations (above 2000 m.a.s.l.); among pottery traders they are called *sierra* or cold (*frío*). Grain sites are found along Trade Routes 1, 2, and 4; but the sectors within the community of Matalacas are the most renowned for grain production. *Recado* sites are located in lower, warm areas, all are below 1600 m.a.s.l.; these are called *temple* or hot (*caliente*) by traders. In Olleros, "*Recado*" refers more specifically to manioc, plantains, and corn on the cob; it does not include highland tubers like potato or olluco, which are generally found at Grain sites. *Recado* crops are grown within the community of Olleros, and so they are mostly found along Trade Route 5. Finally, "Animal" sites are places where it is possible to trade pots for animals including pigs, goats, and chickens. These are mainly the lower, warmer places found along Trade Route 3.

Of these categories of trade destinations, Grain sites are visited most frequently. Grains are sought most often because, while most residents in Olleros have some access to *recado* foods from their own fields, the majority are unable to grow the grains and *menestras* vital to daily food consumption for themselves. Additionally, throughout the Andes there is a much stronger tradition of trading pots for grain than there is of trading pots for other products. In Mohr's (1992) example, for instance, pots were only traded for grain, never for animal products such as cheese or wool. So, we see a combination of a real livelihood need with a deep-rooted exchange tradition that fulfills the need.

Visiting customers also often bring products to Olleros to barter for pots. These customers bring the products of their home communities: those from high altitude grain zones bring grains, *menestras*, or animal products; and those from low altitude, warm zones bring *recados, chancaca*, animal products, coffee, or cacao. Many of the bulk purchasers (*negociantes*) who come from the communities surrounding Ayabaca (especially Suyupampa) are likely to bring "urban" products to exchange for pots. These are products available in the Ayabaca market, including salt, soap, fish, kerosene, or rice. Any customer may also bring money.

The ability of potters to dictate what products and what quality of products they exchange their pots for is an interesting problem. While this was addressed only tangentially in the interviews, several potters did discuss the issue. For example, as Micaela Jiménez explained:

> When one travels to trade, one can only exchange for what they say, "Do you want maize?" Hopefully. "Do you want wheat?" As well. One must accept what they offer...[2]

This issue is also alluded to, albeit vaguely, when potters say they trade for "any kind of thing." Overall it seems that potters have less power in exchange relationships than do their trading partners, but this is not a firm conclusion.

Beyond simply considering what products are traded for and where trade occurs, it is also necessary to think about when crops are available in different places. The annual agricultural cycle and consequent timing of harvests plays a crucial role in exchange practices. As stated in the description of Trade Route 1 to the Community of Matalacas (Sect. 2.2), the sites where pots are traded are scattered across a large area and across a considerable range of elevations. This means that harvests occur over a span of time. Including the maize harvest in Olleros (which marks the start of the potting season), "harvest" time for seasonal crops begins in June and lasts into September, with a special focus on July and August: "the grain months that there are" (Emma Mondragón).[3] This is a result of the range of crops cultivated, but also because growing periods are shorter in lower, warmer parts, while higher, colder places harvest after a longer season. Crops that are available year round (e.g., manioc, sugarcane), or animals/animal products may not present as much of a timing concern. In sum, the most basic feature of the landscape of exchange that traders must negotiate is that different products are available in different places at different times.

3.1.2 Ecology and Agriculture

In the Andes, trading for a variety of agricultural products often means accessing several different ecological zones. This is certainly true among Olleros pottery traders. Since pottery production is not exclusively tied to one ecological zone, the way cultivation of a single crop might be, trade in this manufactured object becomes an interesting perspective from which to discuss questions of verticality and ecological complementarity.

Clearly, the Andes contain a great diversity of environments. At the macro level, these follow fairly predictable patterns based on the variables of longitude, latitude, and elevation (Troll 1968, for a summary see Brush 1977: 2–6). At the micro level they depend on an even wider range of factors, but especially elevation (Brush 1977: 5). The Andes, like any mountain system, exhibit a series of stacked vegetation/life zones which are distinguished by temperature and rainfall levels

---

[2]"Entonces, uno cuando sale a buscar de cambio, uno puede cambiar con lo que le dicen: '¿Quieres maíz?' Ojalá. '¿Quieres trigo?' También. Uno no hay que regodear lo que uno le pronuncian para cambiar" (Micaela Jiménez).

[3]"En Julio y Agosto, Julio, Agosto, por ahí son los meses ya graneros que hay" (Emma Mondragón).

(see Tosi's 1960 classification of life zones, and see Brush 1977 for further discussion of these macro and micro ecological and climatic patterns). Especially significant for understanding the Northern Andes are patterns related to moisture: the heaviest rains fall on the eastern slopes of the Andes, and overall the environment tends to be moister nearer the equator and drier as one moves southwards. These features have consequences for agricultural organization, and thus for understanding (pottery) exchange practices.

More important here than ecological definition of zonation is the way the environment is perceived by its inhabitants and the *lived* zones useful to them. For the Northern Andes, Brush (1977: 80) distinguishes four major crop zones. These are, (in order of increasing elevation): (1) lowland tropical zone for coca, fruit, and sugarcane, (2) temperate grain-producing zone, (3) potato/tuber zone and (4) pasture zone. For the Department of Piura, Hocquenghem and Monzón (1995) provide a detailed description of food production, including crop types and locations of cultivation. They identify four zones, distinguished by elevation and temperature. These are the Cold Sierra (*Sierra fría*), Temperate Sierra (*Sierra templada*), Hot Sierra (*Sierra caliente*), and Coastal Valleys (*Valles costeñas de ríos Chira y Piura*).

Hocquenghem and Monzón (1995) link their zonal definitions to specific elevation ranges and to Tosi's (1960) life zones, but their major contribution is the categorization of ecological zone by the plants and foods most important to rural livelihoods. The following Tables (3.2, 3.3, 3.4, 3.5, and 3.6) provide Hocquenghem and Monzón's (1995) classification of zones and types of products. European grains, Andean tubers, and various legumes are produced at the highest, coldest elevations; maize, tropical tubers and legumes are grown at the middle, warm elevations; maize, tropical tubers, legumes, and vegetables are grown at the low, hot elevations; and, maize, rice, tropical tubers, legumes, and vegetables are

Table 3.2 Piura production zones adapted from Hocquenghem and Monzón (1995: 20)

Region	Elevation range (m.a.s.l.)	Description
Cold Sierra	>2000	Montane Woodland, Páramo
Warm Sierra	1000–2000	High jungle (Selva alta)
Hot Sierra	500–1000	Dry forest (Bosque seco)
Coastal Valleys	0–500	Algarrobo forest (Algarrobal)

Although paramo (high grassland) is found above 3500–4000 m, below that the natural vegetation would be montane woodland

Table 3.3 Grain production zones [translated from Hocquenghem and Monzón (1995): Cuadro 1—Zonas de Producción de los cereales]

Regions	Criollo Maize	Hybrid Maize	Wheat	Rice	Barley
Cold Sierra			X		X
Warm Sierra	X				
Hot Sierra		X			
Coastal Valleys		X		X	

Table 3.4 Tuber production zones [translated from Hocquenghem and Monzón (1995): Cuadro 4—Zonas de Producción de los tubérculos]

Regions	Potato	Oca	Olluco	Sweet potato	Manioc	Arracacha	Achira
Cold Sierra	X	X	X				
Warm Sierra					X	X	X
Hot Sierra				X	X		
Coastal Valleys				X	X		

Table 3.5 Legume production zones [translated from Hocquenghem and Monzón (1995): Cuadro 5—Zonas de Producción de las leguminosas] For specific varieties of beans, see original table

Region	Field peas	Broad beans	Beans	Peanuts
Cold Sierra	X	X		
Warm Sierra	X		X	
Hot Sierra			X	X
Coastal Valleys			X	

Table 3.6 Other cultivars, production zones [translated from Hocquenghem and Monzón (1995): Cuadro 6—Zonas de Producción de otras hortalizas]

Regions	Zambumba	Squash	Caigua	Cabbage	Garlic	Onion	Chili	Tomato	Achiote
Cold Sierra	X			X	X	X			
Warm Sierra	X	X	X			X			
Hot Sierra		X				X	X	X	X
Coastal Valleys		X				X	X	X	X

grown in the coastal valleys. Additionally, there are slight differences in the livestock raised in each zone, with goats and cows especially common in the Hot Sierra.

As stated above, Olleros pottery traders describe three major types of trading sites: Grain, *Recado*, and Animal. These are also classified by their elevation/temperature characteristics: Grain sites are colder and higher, and *Recado* and Animal sites are lower and warmer. These represent the zones that are useful to pottery traders. They are the types of sites where pots can be exchanged for specific products. Brush's categories fit well with the pottery traders' perceptions and with Hocquenghem and Monzón's (1995) definitions. The major difference between Brush's classification and the Piura cases is that because the highest elevations in Piura are relatively low compared to other areas of the Northern Andes, the total area which falls into the pasture zone is limited. Likewise, much of the tuber cultivation in Piura occurs in what could be considered the grain zone in other parts of the Northern Andes. These comparisons are summarized in Table 3.7.

Table 3.7 Comparisons between different crop zones

Pottery traders	Hocquenghem and Monzón (1995: 20)	Brush (1977: 80)	Gade (1975: 104)
	Coastal Valleys: maize, rice, sweet potato, manioc, beans, other vegetables		Manioc and cacao
			Coca and coffee
Temple, hot: Recado sites. Trade for: manioc, plantains, sweet potato, choclo	*Hot Sierra*: maize, sweet potato, manioc, beans, peanuts, other vegetables	*Lowland tropical zone*: coca, fruit and sugarcane	Subtropical starch crops
Sierra, cold: grain sites. Trade for: maize, wheat, field peas, beans	*Warm Sierra*: maize, manioc, arracacha, achira, peas, beans, other vegetables	*Temperate zone*: grain	Maize
Sierra, cold. Grain sites. Trade for: maize, wheat, field peas, beans	*Cold Sierra*: wheat, barley, potato, oca, olluco, peas	Potato/Tuber zone	Wheat, barley, broad beans, potatoes
		Pasture	Native wild grass, occasional tubers

Elevations are not strictly defined here, but elevation increases from top to bottom of the chart
The zones from each scheme have been “matched up” as well as possible
Animal sites are usually found in lower zones, but are not strictly defined by elevation range

These Northern Andean classifications may also be compared with Gade’s (1975: 104) observations from the Vilcanota Valley in Southern Peru (Cuzco), which demonstrate six different ecological regions characterized by the plants useful to local livelihoods. These regions are (from highest to lowest elevation): (1) region of native wild grass and occasionally cultivated tubers; (2) regions of wheat, barley, broad beans, and potatoes; (3) maize region; (4) region of subtropical starch crops; (5) region of coca and coffee; and (6) region of manioc and cacao. Gade (1975: 104) describes considerable overlap between these plant ranges, and overall they exhibit greater complexity than their Northern Peruvian counterparts (see Table 3.7). In all regions crop zones boundaries may be flexible, as they are partly determined by farmer choices, which can be based on relative efficiency of crops, local conditions, market opportunities, and/or personal experiences and preferences (Gade 1996; Zimmerer 1999).

While the ecological variation between regions within the Andes is an interesting topic in its own right, here it is more important to consider how different crop zones are incorporated into rural trade systems. The differences between the pottery traders’ perceptions and the other schemes are partially related to how different crops are exchanged for pots. Traders may be excluding zones they recognize, but where there are no products available for barter. Crops matter to pottery traders because they are a source of food. When asked how pots are traded, typical potter responses included "With maize, olluco they are traded" (Rosa Jiménez).[4] As to why they must travel to places so far away, the answer is even more simple "because here there isn't any" (Orfilia Mondragón).[5] And finally, the situation is succinctly summed up: "Maize, peas, in all the parts of the higher zone, exchange is with grain. (And when it is not high altitude?) When it is lowland, exchange is for manioc, sweet potato, the low parts…*recado*" (Micaela Jiménez).[6] Pottery traders have a clear perception of the landscapes in which they trade their product.

3.1.3 Cuisine and Complementarity

Cultural standards of cuisine determine what kinds of food potters seek to obtain from trade. Weismantel (1988: 87) defines cuisine as the "…cultural construction of meals, the structures that organize knowledge about foods, and the pattern of their preparation and combination." She continues to explain that within a given "cuisine," foods are grouped into categories which fulfill the same basic purpose, but that different foods in the same category often have different symbolic values (e.g., potatoes and rice) (ibid.). Hocquenghem and Monzón (1995) describe the structure of cuisine in Piura. In rural areas of the Piuran sierra, for a main dish to be complete it must consist of three components: (1) Meat/Fish, (2) (Various) *Recados*, and (3) a *Menestra*. The types and variation in *Recados* and *Menestras* consumed by a household can vary with economic and environmental conditions. In hard times the *Menestra* alone can become the main dish (Hocquenghem and Monzón 1995: 43). Rice is also commonly consumed among highland families when they can afford it. Thus there are similarities in how food groups and crop zones are classified. This helps to explain the kinds of foods pottery traders may desire, as well as their perceptions of the environment.

Considering the geography of production outlined above, fulfilling Piuran culinary standards almost always entails accessing products from different production zones, a reality which is, after all, the basis of Murra's (2002 [1972]) verticality theory. For Piura and pottery exchange, a flexible conception of ecological complementarity is probably most useful. It is apparent that pottery traders articulate multiple ecological zones in order to access a range of products through the exchange of one specialized product. This form of complementarity applies to

[4]"Con maicito, olluquito, lo cambiaba. Con cualquier cosita de comidita." (Rosa Jiménez).

[5]"Porque por aquí que no hay pues" (Orfilia Mondragón).

[6]"Maíz, alverja, todo lo que es zona de la parte altura se cambia con grano. (¿Y cuando no es altura?) Cuando es bajera, se cambia por yuca, camote, a la parte baja…el recado" (Micaela Jiménez).

other types of producers as well. In a separate example from Piura (Challe Grande, District of Frias), Apel (1996: 158) found:

> *Campesina* families devote a portion of their harvest for interchange with products that they themselves do not produce. The *comuneros* of Challe only grow small amounts of plantain and sweet potato: they obtain these products through exchange for beans, wheat or peas; they also acquire coffee, chancaca (*dulce*) and cane liquor (*primera*) through barter with grains (my translation).

A similar situation was recorded by Bernex (1990: 100) in her discussion of traders from many points of origin arriving in the grain-producing region of Matalacas with a variety of goods to trade (see to Sect. 2.2).

Thus among *campesinos* in this region there are established trade networks for acquiring goods that one does not produce oneself. These networks incorporate ecological complementarity as well as specialization in production, and rely heavily on nonmonetary or barter exchange of product for product. For the most part they are constituted by informal trade between individual *campesino* households. Kinship may play an important role in these trade relations (Brush 1977), although not always, as in the case of the potters. This system tends to ignore established markets and urban centers, creating independent circuits of exchange. Through these flows of goods—agricultural, animal, and handicrafts—multiple communities become linked in a flexible structure of exchange.

Interestingly, within this system, places become associated with the types of products they grow or make (e.g., *zonas graneras, zonas cabrillas*), and thus the types of products traders can obtain if they travel to that place. This is most apparent by the name Olleros itself, which identifies the presence of potters. In the geographic literature, this identification of product and place is most frequently discussed with respect to geographic indication labeling schemes and the creation of niche markets for foodstuffs produced in traditional territories by traditional methods (Skuras and Dimara 2004; Ilbery et al. 2005). While this literature mainly focuses on evaluating the marketing schemes aimed at promoting the "local" or "territorial" to detached urbanites, it does speak to the cultural embeddedness of food consumption (Skuras and Dimara 2004: 804). Perhaps more significantly, it attempts to show that to promote or conserve the production of specific traditional items, attention must also be paid to the "social and economic structures which sustain them" (Ilbery et al. 2005: 118). Thus far it is apparent that there is a highly specific social and economic context in which Olleros pottery exchange occurs, a point perhaps not considered by the 2004 UNDP ceramic project, but one which could certainly be incorporated in future plans.

While these can scarcely be called new findings—after all, ecological complementarity is one of the most discussed themes in the Andeanist literature—this study's major contribution is through analysis of complementarity from the perspective of trade in handicrafts. In fact, there are several important insights that can be derived from pottery trade that are more difficult to see when the focus is only on food exchange. These are addressed in the following discussion of exchange rates and measurement systems.

3.2 Exchange Rates and Methods of Measurement

3.2.1 Exchange Rates

To this point, methods of pottery distribution, trade routes followed and customer points of origin, and products exchanged have all been discussed, however, how pots are actually exchanged, rates of exchange, and the measurement systems that are used are topics that have been purposefully neglected. This is an important issue warranting its own focus. In the Andes pots are "commonly bartered by filling the vessel with the produce (such as grains, beans, tubers) for which it is being exchanged" (Sillar 2000: 83). Antonio Raimondi made one of the earlier observations of such exchange: "In the latter part of the last century he found the earthenware storage and cooking vessels (*ollas*) made in Junín exchanged against the measure of wheat they could hold" (Wrigley 1919: 73, referencing Raimondi 1873). Pots have also been used as measures for exchanging other products. Burchard (1974: 234) refers to a "*tupu* or *poto* of clay used for cooking but also as a unit of measure" (My translation).[7] Fonseca (1973: 123) and Mayer (2002: 153) likewise mention measurement using ceramic cooking pots.

Sillar (2000: 83) argues that "the use of pottery as a measure is almost certainly a pre-Hispanic custom." He cites the Quechua concept of *pokcha*: defined in Holguín (1989 [1608], 291) as a "measuring vessel of half a fanega's capacity," as well as the Aymara concept of *tanca vicchi* defined in Bertonio (1984 [1612]) as filling a clay jar up to the mouth with maize or other food. Rostworowski (1960) provides similar explanations of these two terms.[8] Clay jars were also used as measuring devices in colonial Spain, the *arroba* was a measure of weight but also of capacity, it referred to the standard terra cotta jar (*amphora*) used for storing liquids such as wine or oil.[9]

Regardless of the origin, the custom of exchanging a clay vessel for its fill of produce is common throughout the Andes. Pottery in Olleros is most often bartered within this capacity system, termed "*a la llenada*" (by the potful) in Piura and elsewhere. Within the *llenada* (potful) system the pot can be filled to varying levels, which are sometimes referred to by the morphological features of the pot: half (*mitad*), shoulder (*hombro, hombrito*), neck (*gollete, golletito*), and full (*llena*), which may be either striked (*corrida de vara*) or heaped (*flotando*). For Southern

---

[7]Sillar (2000: 85 n. 34) writes that while *tupu* is a word commonly used as a unit of measure, he has never heard of it in reference to a pottery vessel. In Ecuadorian Kichwa, the verb *tupuna* means "to measure." In Piura, a *poto* is a gourd used as a vessel (bowl, cup, etc.), I have never heard it used to describe a pot.

[8]A fanega is volumetric measure: "A unit used in Spain and Portugal (and therefore sometimes in South American countries…) in both liquid and dry measure, although the Spanish variant is applied predominantly to dry goods. The unit differs by a small margin, however between the two traditions," it is about 55.5 L (Darton and Clark 1994: 147).

[9]An *arroba* is also a measurement of weight, it varies slightly depending on the region, but it is equivalent to about 11.5 kg (*Diccionario de la Real Academia Española*, s.v. "arroba").

Peru and Bolivia, Sillar (2000: 85) lists the handles, neck, rim, and decorative line drawn around the neck as features used in measurement. He also alerts us to the fact that when making pots, potters pay close attention to where these features are placed. For example, handles are not placed too low on the pot's body, because this would result in lower yields of produce when the pot is bartered.

The level the pot is filled to in exchange reflects many factors, including, but not limited to: the product being exchanged, the customary equivalences recognized by potters and buyers for that product, the type of pot being traded, the location where trade occurs (place of production or consumption), and the season (during, before, or after harvest). While several of these themes have already been discussed with respect to the geography of agricultural production, here they will be analyzed for their influence on rates of exchange in barter transactions (see Appendix D for a registry of products and exchange rates).

The accepted levels to which the pot should be filled vary for different products, but are well known to both potters and buyers before the exchange occurs. As stated above, the most common products exchanged are (in order): field peas, maize, wheat, and beans. Of these, pots are always filled full with maize and wheat, but can be filled anywhere from half to full with peas and beans. This mostly depends on where the exchange occurs. Potatoes, broad beans, and coffee beans are also traded with the *llenada* system, but their levels were unspecified. Olluco is the only product said to be exchanged "heaped" (*flotando*) (Mauro Mondragón).

Additionally, the form of the pot being traded matters for how it is exchanged. Regular pots (*ollas* and *sapimas*, see Fig. 1.2) are worth less than toasting pots (*tiestos*). This means that for the same product (e.g., beans) a regular pot would be filled half-full, when at the same time and place a toasting pot would be filled completely full. This difference in value refers to the greater difficulty in transporting *tiestos* than *ollas*, *tiestos* are more likely to break en route, and so are a rarer commodity. *Tiestos* are also more valuable because they are less commonly produced, and their toasting function cannot be easily replaced by a metal counterpart. Given the difficulty in transporting pots on the backs of donkeys, exchange rates favor the traveler. Potters traveling to distant communities, (especially Matalacas, the farthest place from Olleros), have their pots filled to higher levels than they do when traders bring grain to Olleros to buy pots.

Barter exchange rates are said to be unchanging, they have been the same as long as the most senior generation alive today can remember, and apparently are not affected by changes in the national currency value (Mauro Mondragón). Continuity in exchange rates indicates that the value of clay pots is not influenced by the influx of ever more, and ever cheaper, plastic, and metal vessels, although the number of clay pots being produced almost certainly has declined. This implies that clay pots move within a separate circuit of exchange than the industrial product; and that rates are determined by relationships between individuals, instead of by more formal economic logic. Change in pottery production and exchange over time is further discussed in Chap. 4.

The seasonality of trade also affects exchange rate. Potters almost always trade between the months of July and November, with a peak in exchange during August.

As described earlier, this corresponds to both the best time for traveling and for firing pots. Even more compellingly, it corresponds with the timing of harvests in the communities where potters trade. Trading during and just after harvest guarantees a better exchange rate. Mohr (1992: 76) suspected that "it is likely that more produce is obtained for a pot after harvest when supply of produce is higher," but never was able to observe this pattern. At least in Olleros, this does indeed occur, as one potter, Micaela Jiménez, explained, "When it is during harvest, for full pots, but when it is in *less* harvest, then filled a little lower."[10] Others have likewise observed higher exchange rates for bartered products during and just after harvests (Gade 1975: 52).

Exchange rates follow identifiable patterns along each of the five trade routes. (see Table 3.8 for exchange rate data). In Matalacas (Route 1) potters can get the highest quantities of peas, maize, wheat, and beans for their pots (full potfuls). In the other Grain zones (Routes 2 and 4), potters generally get full potfuls of maize and wheat, but less (as little as half potfuls) of peas and beans. Chickens are also traded at a better rate at the Route 2 sites than at the Route 3 sites. These patterns help to explain why potters travel longer, more difficult routes to acquire the same products that can be found much closer to home.

And finally, the value of a pot increases as it is traded further and further from the production center, outwards along the networks of exchange. Often, as described above, during trading trips potters will trade with customers who purchase more pots than they plan to use in their individual household. These customers then re-trade the extra pots to other consumers even farther afield. In these cases, it is said that exchange rates can reach as high as two potfuls of grain for each pot. The potter Micaela Jiménez most clearly described this situation:

> No, here [in Olleros] there is no trade for two potfuls. A long time ago, it is said that they traded for two potfuls, but one had to go farther, to where no other pots enter, only there do they trade for two potfuls. Because, for example, there are places where people buy lots of pots, and other people from even farther know that these people buy pots, and so then they say that they give two potfuls. In Matalacas they buy a lot of pots from me, one *carga*, two *cargas*, and then they trade them like that... they save a few for themselves, and people from farther, from Pedregal, from Huaracas, which are located over there, up that slope where we don't go, come down to them for the rest [of the pots]... [They are] from Huaracas, from Matalacas, but to get to Huaracas you have to go veeeeerry far into that sierra.[11]

---

[10]"Cuando está en buena cosecha, a la llenadita, pero cuando está en menos cosecha, ya un poquito más abajito" (Micaela Jiménez). Her use of the phrase "buena cosecha" could mean "good harvest," implying a year with an extra abundance of quality produce or "during harvest," implying while the seasonal activity is occurring. I am leaning toward the second definition, mainly because the explanations of other potters tend to coincide.

[11]"No, acá no se cambian a dos llenadas. Más antes si decían que cambiaban por dos llenadas. Pero hay que irse más lejos, por donde no entran no más olla- ahí recién cambian a dos llenadas. Porque por ejemplo hay sitios donde compran bastante olla y la gente de más lejos ya saben que allí señores compran, allí dicen que les dan a dos llenadas. A mismo Matalacas, me compran bastante, una carga, dos cargas, y así las cambian...hacen quedar un poco para ellos, y lo demás ya viene gente de más lejos de Pedregal, de Huaracas que se quedan para acá...nosotros no vamos por esta hondura así...de Huaracas, de Matalacas, pero para Huaracas pasa lejísisimos a la sierra en esta sierra" (Micaela Jiménez).

Table 3.8 Exchange rates for the four most commonly traded products, as well as for a commonly traded animal product at sites along the five trade routes

Route	1. Matalacas	2. Passing el Aypate	3. Lowlands to South	4. Passing Ayabaca	5. Within Olleros
Peas	Full potfuls	Half potfuls	n/a	n/a	n/a
Maize	Full potfuls	Full potfuls, a little less if not in harvest	n/a	Full potfuls	n/a
Wheat	Full potfuls	?	n/a	Full potfuls	n/a
Beans	Full potfuls	Half potfuls	n/a	n/a	n/a
Chickens/ Hens	?	1 hen = 1 regular pot	1 hen = 2 pots	n/a	?

About annotation "*n/a*" means the product is not available along that trade route
"*?*" means the product is traded, but the exchange rate was not recorded (see Appendix D for a complete registry of products, places, and exchange rates)

This example illustrates how pots continue to move through exchange networks after they are traded by their producers; but also, that the farther pots move outwards from the point of production, the more valuable they get. Even so, there is a chance that the two-potful exchange rate is an exaggeration, or an almost mythical understanding of what actually happens. It seems unlikely that the potters describing these exchanges have ever actually witnessed one.

Additionally, no traders mentioned haggling, but they were also not explicitly asked about this aspect of trade. In other Andean regions, simply knowing customary exchange rates does not rule out active bargaining. Sillar (2000: 85) cites the importance of both haggling and giving a *yapa*, or little extra bit of produce, for successfully sealing the deal during exchange. Mohr (1992: 77) also reports that potters are particular about the quality of the products they exchange pots for, and may refuse to trade if the product is not acceptable.

Pots can also be exchanged for a set weight or count of a product. Exchange by weight is more often used for products whose shape or value does not readily correspond to pot volumes (such as wool or manioc), or for products which are less commonly traded (such as cacao). Animals are often traded by the count or unit; for example a goat is said to be worth approximately 12 *ollas*, although meat can also be traded by weight. Pots are more commonly traded for chickens than goats or goat meat, one henchicken is equivalent to one regular sized pot. Cheeses are also traded for pots, one cheese can be worth from 1 to 2 pots. These exchanges should be contrasted with Mohr's (1992) observation in Cuzco, that only grains are traded for pots, not meat or any other food or material. In general, exchange based on weight or count (unit) seems to be more flexible and thus more difficult for an outsider to comprehend.

Finally, pots from Olleros are also sold for money. While the price of a regular pot (*olla*) is said to vary, it is at least 2 *soles* (approximately $0.65 at 2006 values). Potters all mention that they sell "for money" ("*por la plata*"), but do not give the detailed descriptions they do for exchanging pots for food. Sillar (2000: 85) gives one example in which a pot's monetary value is calculated by working out how much the amount of maize needed to fill it would cost.

3.2.2 Measurement Systems

Recognizing that there are a diversity of exchange methods and measurement systems used in pottery distribution, how can we make sense of them and what can they tell us about artisans' livelihoods? This is the area where analysis of trade in material objects yields especially important insights. To begin, we recognize that measures of capacity are by far the most common method used for trading pots. Measures of capacity have long been used in many different contexts. With regards to European history:

> Many goods that we are now accustomed to buy by weight used to be purchased by customary measures of capacity or by the piece; thus, cheese was sold by the chunk or slice, butter by the 'round', wool by the fleece...throughout Europe, in medieval as well as in modern pre-metric times, (and even later, here and there), both liquids and dry goods were measured by volume or measures of capacity, such as the bushel, the gallon...(Kula 1986: 42).

Measures of capacity were also used in the pre-Hispanic Andes, the *pockha* and *tanca vicchi* are examples of such measures described above. These "folk measures" are thought to have derived from aspects of the human body; including the hollow of the hand, the handful, the armful, and the load of the man (Tuan 1977: 45–6). They then developed further out of "the conditions, objectives and outcomes of human labor" (Kula 1986: 5), and especially from methods of transport, such as the basket, sack, wagonload, boatload, and cartful (Kula 1986: 5; Cronon 1991: 107). Beyond simply being practical ways for dealing with the logistics of exchange, capacity measures make sense conceptually. The qualitative differences between products lead to the use of different measurements for different products (Kula 1986: 69).

Even once capacity measurements are agreed upon, conflicts still arise over the way that the measurement is physically carried out. For grain measurement in medieval Europe, Kula (1986) demonstrates the great importance of practices like whether the grain is "striked" or "heaped" when poured, the diameter of the capacity measure used (and thus the volume of the heap), and how well the grain is packed. He describes many widespread patterns in measurement systems which are also present in the pottery example, including: one method of measurement is used before the harvest and another afterwards, and different measures are used in the places of consumption and production in order to cover the cost of transport. Thus the *method* of measurement can have great significance for the real amounts of grain exchanged. In Medieval Europe, strategic use of different methods often amounted to the profit made by a merchant (i.e., buy grain "heaped" and sell it "striked"), or the interest earned by feudal lords on loans to peasants (i.e., lend grain "striked" and have it returned "heaped") (Kula 1986). Given these circumstances, it is not difficult to imagine how abuses in measurement may be perpetrated, and why these practices have often been fiercely contested.

As we see, pottery exchange in the Andes fits many of the general descriptions of measurement by capacity, with the important distinction that one of the products being exchanged is also used as the measure. Fonseca (1973: 123) cuts to the heart of the issue: "They use the actual pots (*mancas*) and blankets (*mantas*) as measurements of exchange, in no case do they refer to 'prices,' since the same objects being exchanged also serve as measures" (my translation). However, it seems that more than just convenience in exchange is happening here. The key is in the object itself and the measuring system that the object implies. As described above, potters are not the only ones who trade pots made in Olleros. Many other people buy pots from potters, paying in cash and kind, to transport them to other places to trade. The idea is that, as pottery trader Altagracia Chuquihuanca puts it, "Trading pots results better….because they give potfuls, by the pot."[12] Bartering pots instead of buying products with money gives the trader better returns, because pots are traded at the more advantageous rate of "*a la llenada.*" Given Kula's (1986) explanations of the patterns that result under capacity measurement-based exchange systems, it is possible to understand how pots can be used to get "better results" depending on where and how trade occurs.

Earning a profit through secondary exchange of products has been well documented in the Andes. Mohr (1992: 87) observed that "produce bought or obtained in barter may be used to exchange for other items, as can cash to make purchases." Others, particularly Mayer (2002: 151), describe this practice well: the basic idea is that traders strategically move from place to place, including through different ecological zones, "taking advantage of differential exchange rates between cash and barter transactions, as well as absorbing costs from field to road." When understandings of manipulation of measurement in exchange are combined with discussions of trade between ecological zones, it is not hard to see how individuals can make use of this overall geography of production and exchange to earn a living.

Additionally, and most importantly, Kula (1986) alerts us to the social significance of differences in methods of measurement, pointing to conflicts over measurement as representative of class struggle and diverging interests between members of society. The social pressures to continue using one measurement system can be great. In Nepal, Humphrey (1985: 66) found that the "absence of exact measurement is not a feature of low cultural development but a deliberate

[12]"Bueno, con la plata también pero aquí a veces no hay—mejor lleva su… mejor resulta el cambio con sus ollitas. Resulta mejor." (¿Como así le resulta mejor?) Porque ellos dan a la llenada, según la olla. (Altagracia Chuquihuanca).

strategy to ensure the persistence of one-to-one transactions." In Bolivia, (Sipe Sipe, Cochabamba), Delgado and Ponce (2002: 8) describe exchange of agricultural products with baskets and special bags called *chimpu*, as a way to incorporate individual needs, harvest quality, and "the affection or friendship between one and the other." Measurement and exchange methods are controlled by social relationships between individuals. This may be a factor in the stability and persistence of rates of exchange for pots.

The power relations involved in exchange may also be revealed through analysis of measurement systems. While this is not often addressed in the Andean literature, Göbel (1998: 885) strongly alludes to such a power differential between herders and farmers in northwestern Argentina. In recent years, the farmers have switched from measuring grain in "traditional" capacity measures like the *raso* or the *almud*, and now exchange in buckets, baskets, and boxes of varying sizes. The herders are forced to accept these measures, although with much grumbling, and with threats to switch the entire system to measurement by weight (called *rumana*). Similar tensions between potters and their customers may also exist; this is an area where more research is needed.

3.3 Potters' Life Cycles and Trade Practices

The trade patterns discussed so far: agricultural organization, products and places, exchange rates and measurement systems, are all important features of the exchange landscape which traders must negotiate. However, it is important to understand that this negotiation occurs at the individual and/or household level. Each potter, and/or her family, makes decisions about the marketing of her pots based on her/their own needs and abilities. In other words, pottery production is a capacity a household can draw upon as desired. This may relate to how much she relies on pots for income ("assets" according to Reardon and Vosti 1995), or factors even farther from her control related to climate, the market economy, or other events. However, even though pottery production is based at the individual household, we may identify several overarching trade patterns relating to a "typical" potter's life cycle.

A potter's stage of life greatly influences how her pots are traded and distributed. As an unmarried girl, the potter may sell her own pots from home, but her father/family are responsible for taking the pots on trading trips. When the potter starts her own household (either by growing up, marrying, or becoming a single mother), she begins to take her pots to trade by herself or with her husband, friends, or children. As the potter ages, her trading area becomes more constrained, she travels to places more easily reached, either because of the distance or difficulty of the route. Her husband or other family members may still take the pots to farther locations. Of course there are many exceptions to this pattern, for instance: some potters never/rarely go on trading trips at all, waiting for visiting customers instead. Another potter stopped making pots after marriage, because her husband, a descendent of the ex-landlords did not like the activity (see Chap. 4 for the sociocultural connotations of pottery).

While pottery trade decisions represent an individual's needs, experience, and resourcefulness, there are general patterns shared across the community. These patterns likely shift over time. For example, as fewer and fewer girls are currently learning to make pots, pottery activity is becoming constrained to the middle-aged and elderly generations. Chapter 4 more fully discusses such long-term shifts.

3.4 Conclusions

Returning to Polanyi's classic description of trade, several final reflections can be noted. Polanyi (1975: 144) writes that:

> The decision to acquire some kind of goods from a definite distance and region is necessarily made under concrete circumstances different from those under which some other kind of goods would have to be acquired from somewhere else…trading ventures differ according to the type of goods to be acquired and transported, thus forming separate branches of trade, each with its distinctive operational methods and organization.

The decisions and negotiations necessary to successfully trade pots in Olleros and elsewhere incorporate a range of features, which include regional agricultural patterns, types of products sought, measurement and exchange systems, individual or household capacities, and social relations at individual and community scales. The spatiality of pottery exchange reflects these exchange decisions, or in other words, depends on the way the potter or trader negotiates the exchange landscape. Pottery exchange is only ever one component of a potter household's overall livelihood strategy; this broader strategy has its own spatiality that incorporates factors beyond those listed above. Finally, pottery exchange networks are unique from networks of exchange of other goods, and pottery trade in different communities can be expected to follow different patterns.

References

Apel K (1996) De la hacienda a la comunidad: la sierra de Piura 1934–1990. IEP/IFEA/CNRS, Lima

Bernex N (1990) El distrito de Pacaipampa: unidad y diversidad. In Bernex N, Córdova H, Roux JC (eds) El proyecto de desarrollo rural integral de la sierra central del departamento de Piura (Convenio PUC-ORSTOM) Tercera parte-Aspectos socio-económicos. PUCP, Lima, pp 88–103

Bertonio PL (1984[1612]) Vocabulario de la lengua Aymara. CERES, Cochabamba

Brush SB (1977) Mountain, field, and family: the economy and human ecology of an Andean valley. University of Pennsylvania Press, Philadelphia

Burchard RE (1974) Coca y trueque de alimentos. In: Alberti G, Mayer E (eds) Reciprocidad e intercambio en los Andes peruanos. IEP, Lima, pp 209–251

Cronon W (1991) Nature's metropolis. Norton, New York

Darton M, Clark JOE (1994) The dent dictionary of measurement. J. M. Dent, London

Delgado F, Ponce D (2002) La Feria—Fiesta de Siete Viernes. Revista Compas 5:7–9

Diccionario de la Real Academia Española (22nd ed) (2004) Accessed at: www.rae.es

Fonseca C (1973) Sistemas económicos andinos. Biblioteca Andina, Lima

Gade D (1975) Plants, man and the land in the Vilcanota valley of Peru. Dr. W. Junk B.V. Publishers, The Hague

Gade D (1996) Carl Troll on nature and culture in the Andes. Erdkunde 50:301–316

Göbel B (1998) 'Salir de viaje' Producción pastoril e intercambio económico en el noroeste argentino. In Dedenbach-Salazar Saenz S, Arellano Hoffmann C, Köning C, Prümers H (eds) 50 años de estudios americanistas en la universidad de Bonn—Nuevas contribuciones a la arqueología, etnohistoria, etnolingüística y etnográfica de las Américas, vol 30. BAS, Bonn, pp 867–891

Hocquenghem A-M, Monzón S (1995) La cocina Piurana. CNRS/IFEA/IEP, Lima

Holguín DG (1989 [1608]) Vocabulario de la lengua general de todo el Peru llamada Lengua Quechua o del Inca. UNMSM, Lima

Humphrey C (1985) Barter and economic disintegration. Man 20(1):48–72

Ilbery B, Morris C, Buller H, Maye D, Kneafsey M (2005) Product, process and place: an examination of food marketing and labeling schemes in Europe and North America. Eur Urban Reg Stud 12(2):116–132

Kula W (1986) Measures and men. Princeton University Press, Princeton

Mayer E (2002) The articulated peasant: household economies in the Andes. Westview Press, Boulder

Mohr KL (1992) The organization of production and distribution of traditional pottery in South Highland Peru. In: Bey GJ III, Pool CA (eds) Ceramic production and distribution. Westview Press, Boulder, pp 49–92

Murra JV (2002 [1972]) El control vertical de un máximo de pisos ecológicos en la economía de las sociedades andinas. In Murra JV (ed) El Mundo Andino. PUCP Fondo Editorial/IEP, Lima, pp 85–125

Polanyi K (1975) Traders and trade. In: Sabloff JA, Lamberg-Karlovsky CC (eds) Ancient cvilization and trade. University of New Mexico Press, Albuquerque, pp 133–154

Reardon T, Vosti SA (1995) Links between rural poverty and the environment in developing countries: asset categories and investment poverty. World Dev 23:1495–1506

Rostworowski M (1960) Pesos y medidas en el Peru pre-hispanico. Imprenta Minerva, Lima

Sillar B (2000) Shaping culture: making pots and constructing households. An ethnoarchaeological study of pottery production, trade and use in the Andes, vol 83. British Archaeological Reports, Oxford

Skuras D, Dimara E (2004) Regional image and the consumption of regionally denominated products. Urban Stud 41(4):801–815

Tosi JA (1960) Zonas de vida natural en el Peru. Instituto Interamericano de Ciencias Agricolas, Lima

Troll C (1968) The cordilleras of the tropical Americas. Aspects of climatic, phytogeographical and agrarian ecology. In: Troll C (ed) Geo-ecology of the mountainous regions of the tropical Americas. Ferd. Dümmlers Verlag/UNESCO, Bonn

Tuan Y-F (1977) Space and place. University of Minnesota Press, Minneapolis

Weismantel MJ (1988) Food, gender and poverty in the Ecuadorian Andes. University of Pennsylvania Press, Philadelphia

Wrigley GM (1919). Fairs of the Central Andes. Geographical Review 7(2): 65–80

Zimmerer KS (1999) Overlapping patchworks of mountain agriculture in Peru and Bolivia: toward a regional-global landscape model. Hum Ecol 27(1):135–165

Chapter 4
Pottery Exchange and Livelihoods: An Assessment

Abstract This concluding chapter assesses pottery production and exchange as a livelihood strategy. It looks at the ecological, social, and economic motivations for the continued practice of pottery exchange. It offers an interpretation on how pottery practices have changed over time, as well as potential trajectories for the future of this craft. Finally, it indicates the broader significance of this study in terms of understanding livelihood diversification and the spatiality of livelihoods.

Keywords Household · Livelihood · Women · Handicraft · Pottery · Andes

So why make pots? Why trade them? Is it to fulfill cultural culinary requirements through ecological complementarity? Is this selective engagement with the non-monetary economy (barter) an act of resistance to the market through a recourse to a more traditional moral economy? Or, is pottery a reaction to land scarcity and poverty, and thus a symptom of unequal integration into the market economy? Perhaps it is a combination of all of these factors. This concluding section assesses pottery production and exchange as a livelihood strategy, looking at the ecological, social, and economic motivations for its continued use. It also offers an interpretation on how pottery practices have changed over time, as well as potential trajectories for the future of this craft.

Most simply, pots are a useful livelihood activity for poor households. At least in Olleros, potters do not need money to make pots. As explained in Chap. 1, all of the raw materials are available to community members without charge. Pots may then be transformed into food through trade, by methods described in Chaps. 2 and 3. Regional cuisine standards control what foods are desired; and, as discussed in Chap. 3, acquiring these foods entails accessing a range of ecological zones. But the main point is that through barter, potters do not need money or farmland to supply their households with (at least some) food. Pottery production can thus provide a certain level of security to households low on other resources. While pottery is not a direct result of poverty, it is an important skill or capacity which Olleros residents can call upon when they choose.

M.G. Bell, *Pottery, Livelihoods, and Landscapes*, SpringerBriefs in Latin American Studies, DOI 10.1007/978-3-319-52331-6_4

The issue of physical scarcity of cash certainly plays a prominent role in pottery exchange. Perhaps this is most easily seen among the Olleros residents who trade pots but who do not produce them. The trader who remarked that trade with pots gets "better results" (see Sect. 3.2), prefaced her statement with "[I buy products] with money too, but sometimes there isn't any—better to take—Trading pots results better" (Altagracia Chuquihuanca). Altagracia covers her reluctance to discuss a lack of money with the explanation of the benefits of trading pots. Even though she may lack cash, Altagracia is able to trade items she herself produces, especially textiles, with her neighbors for pots. She can then take these pots to other communities to trade for food. Note that she cannot take her textiles to other places to trade, she must take pots. Pots, as material objects that are not produced just anywhere, as textiles are in the Piura highlands, are thus endowed with some of the characteristics of currency. In Peru similar processes have been observed in other places. For instance, in the Community of Manchiri, (Ayacucho, Southern Peru), Valle (1971: 168) found that:

> …they know the national currency, but it has a very limited use among the population; in the past they used eggs and chili peppers instead of money. They bartered with the neighboring population of Saqsamarka, exchanging food products for ceramic pieces (my translation).

Why certain objects gain such characteristics while others do not is an interesting question, and one that warrants further investigation. Yet all evidence points to the inherent value of pottery in this particular exchange setting, further confirming pottery-making as a useful skill and pottery trade as an important livelihood activity.

More generally, many similar types of small-scale commerce are commonly practiced among land-poor Andean households (Flores Ochoa and Nájar Vizcarra 1976). Peddling of goods may be understood as one aspect of a diversified livelihood strategy. It is especially useful for households with limited agricultural assets (Reardon and Vosti 1995). These traders are able to tap into a flexible framework of exchange, which is structured by aspects of ecological complementarity and specialization of production, to supply themselves and their families. This means that traders work within a geography of production (agricultural, pastoral, and handicraft) in which different products are available in different places at different times. As discussed above (Sect. 3.1), this also results in places becoming associated with the items they produce. In Olleros, potters' and traders' perceptions of their surrounding region are closely tied to livelihood activities.

This also means that many households in the Andes acquire the goods important to their daily lives from beyond their home communities, through links with other households in other communities (c.f. Orlove and Custred 1980: 42; Brush 1980: 221). Of course, this is a pattern with deep historic roots. Such barter has been observed in the Andes since the early colonial period (Larson 1995, 1998), and has adapted to and been adopted in a range of situations since that time. To more fully understand the flows of goods which are used to sustain rural highland livelihoods, a multi-community scalar approach is required, and trade networks are an excellent analytic focus for beginning to identify and visualize intercommunity linkages (McSweeney 2004).

Perhaps tangentially to this project, but of extraordinary relevance to archaeologists and ethnoarchaeologists interested in pottery production, is this detail that pots are produced to be traded beyond the potter's home community. Understanding pottery exchange networks nowadays can provide insight into pottery production and exchange and intercommunity relations in Peru in the colonial and pre-Hispanic past. The implications of this study for such archaeological ends are discussed in more detail in Ramón and Bell (2013).

More significantly for the purposes of this book, is that this flexible framework of exchange is also structured by selective engagement with both the nonmonetary sphere (through reciprocity or barter) and the monetary economy (through use of cash and trade in commodities). In Olleros, the nonmonetary sphere is used daily by community members. Exchanges in this sphere take the form of both small-scale reciprocal borrowing relationships between neighbors, and barter of goods with traders from more distant communities. Here, the use of nonmonetary exchange systems allows traders to gain some of the protection described by Mayer (2002: 157–8):

> …barter networks constituted an economic sphere that was separated from the cash sphere, constructed and maintained by the peasantry who, for their own purposes and to their own advantages, tried to isolate it from the cash nexus. Barter in this region was a form of protectionism. Peasants isolated a flow of goods to favor a self-selected group of other peasants, who used cultural norms to create a **separate circuit of goods for their own benefit**. (my emphasis).

These alternative circuits are conceptually separate from the conventional economy and they have their own, distinct spatialities.

A second example, in a similar spirit, is described by Salomon, albeit for an entirely different context than Mayer's case. Here, describing the connections between highland people near Quito, Ecuador and the "yumbos" or jungle inhabitants:

> The little aboriginal communities ringing Quito have for a least five centuries been in both cultural orientation and economic practice, more thoroughly trans-Andean than any of the imperial civilizations which tried to unify forest and Sierra in a state-centralized scheme of integration. Pantropical cosmopolitans of a multitiered multi-ethnic landscape, the Quito Runa have been and still are the cultural switchboard and **economic depot of a trans-montane integration unknown to state planners**. (Salomon 1981: 192, my emphasis).

The point here is that often there are established traditions of exchange and transport of goods which effectively avoid urban centers and marketplaces. This rural interconnectivity may not be immediately apparent to outsiders, especially development practitioners, and the networks of exchange are not likely to follow the predictions of rational economists (Turner and Williams 2002). Qualitative and ethnographic study of these local practices is thus important.

These traditions may be interpreted as acts of resistance to imperialism of both political and economic powers, through recourse to nonmonetary exchange customs.

Mayer and Salomon certainly imply that these exchange networks are purposefully kept separate for the benefit of their users (see also Larson 1998 on the rival peasant economy). These practices represent one of the solutions that Andean peoples have found for maintaining an adequate level of stability in their uncertain and variable agricultural and economic world (Mayer 2002). Such trade is a livelihood capacity; it is a type of knowledge that can provide security as needed. This interpretation was implied in pottery trader Mauro Mondragón's explanation of exchange practices during the Hacienda period. When asked if potters owed any of their pottery income to the landlord he responded that it was all "for us, they didn't understand that."[1] Here, in basic terms, he described both the separateness, as well as the utility, of pottery exchange practices. This freedom, or independence, is an appealing feature of barter, and another reason why such exchange continues to be important.

However, using pots instead of currency and working within the barter sphere may be both a blessing and a curse. Humphrey (1985: 67–8) presents the much less optimistic view of the features of barter described above: "Barter is a response to increasing poverty on the part of people who wish to maintain their autonomy." In Olleros, while it is quite possible to make a living producing pots, it is much more difficult to use pottery production to accumulate the assets that are the real markers of wealth: land, a *trapiche* (for grinding sugarcane), and livestock (especially bulls). While separation from the cash sphere helps potters to provide for their households, it is a somewhat confining system. No matter how much a potter produces, she is likely to remain in the same stagnant position. This is the exact situation that Reardon et al. (2001: 403) found for households "pushed" into nonfarm activity in order to survive. Pottery is an important capacity Olleros women may draw upon, but, at least nowadays, it is on the level of a coping strategy or means of survival (see Kay 2008).

Additionally, this selective engagement in the market carries social and racial significance. As Ortiz (1980: 139) puts it:

> A style of life that relies on purchased imported goods and foods serves to symbolize basic economic distinction between Indians and whites. The former are subsistence producers who also sell cash crops; the latter are totally involved in production for the cash market.

This same relationship was seen in Weismantel's (1988) discussion of the consumption of different food types, (e.g., potatoes v. white rice::home-grown v. store-bought), as symbolic of an individual's racial self-affiliation. Pots themselves are not free of these meanings. According to Orlove (1998: 209), they are earth touching objects "closely bound up with many aspects of Indian life." In Olleros, such connotations of "Indian-ness" are not immediately apparent. However, this undercurrent could be perceived when one potter gave up her craft because her husband, a descendent of the ex-landlords, did not like how the work looked. But overall, pottery production in Olleros can more safely be interpreted as a poor

[1](¿Todo eso era para uds.?) "Para nosotros, ellos no entendían eso" (Mauro Mondragón).

woman's activity than as an indigenous one: the women in Olleros with fewest resources are more likely to make pots.

While it is clear that Olleros pottery production has not been immune to the forces of time, tracing its change is difficult. When asked about the history of their craft, Olleros potters declare that it was given to them by their ancestors: "the first ones, the old ones more than any" (Orfilia Mondragón).[2] The community's name, "Olleros," implies that potters have been present here at least since the seventeenth century, when it was given its Spanish name (see also Sect. 1.4). It is clear that pottery production has been important to women's livelihoods for generations. During Hacienda times it provided extra relief to single mothers who, although they lacked spouses, still owed the same amount of labor to the landlords (Daniel Mondragón). Currently, Olleros can boast one of the most intensive levels of highland pottery production in Northern Peru, yet it is also true that fewer young women are practicing this craft now than in previous generations, and fewer potters are teaching their daughters.

Here and in other communities with potters, in recent history (the past 20 years or so) there has been an observable decline in production, much of which has been attributed to the influx of cheap metal and plastic vessels. The impacts of these new products on aspects of pottery exchange beyond the quantity of pots traded, such as pot value and exchange rates, or relationships between trading partners, are unclear. The reality that fewer visiting traders and middlemen and women are soliciting Olleros potters is illustrative of a declining or changing market. That potters are more likely to exchange their pots on trading trips than in their own homes indicates the asymmetrical balance of their existing exchange relations. But as this book has tried to demonstrate, pottery can still be used to meet specific livelihood needs, and so I predict that it will continue to be produced.

Pottery can only fulfill this function within the broader barter exchange context in which it is now embedded. The tradition of exchanging pots for potfuls of grain is longstanding. This market for pots will persist, as cash scarcity in the Andes is not unique to communities with potters. Specifically, this means that many rural households may be more likely or able to barter farm products for clay pots than they are to purchase expensive metal pots. We see that this barter system addresses many of the daily needs of rural highland peoples, ranging from the material (food) to the more intangible (autonomy). Rural exchange networks are bound up in all aspects of highland life, and consequently may be read for many economic, political, and sociocultural features of this life. The UNDP ceramic project, which sought to be "an alternative of sustainable economic development…exercised by women, who, with their work develop an efficient use of their natural resources and [foment] an appreciation for women in the economic and social development of the village" (UNDP et al. 2004, my translation), was not totally wrong. Pottery is a sustainable economic activity, practiced by women, which makes efficient use of natural resources, just not via the cash-centered approach planned by the project.

[2]"…las primeras, las antiguas más que todo" (Orfilia Mondragón).

Pottery as a livelihood activity thus speaks to a range of issues facing rural Andean people. The previous chapters have focused on understanding the landscape of exchange within which pottery trade networks are created and sustained. They have shown how pottery traders negotiate the geography of agriculture, seasonal cycles of production, exchange rates, and measurement systems. However, pottery as a livelihood activity also speaks to the broader literature on livelihood diversification and the importance of nonagricultural or off-farm activities to rural households. Balancing multiple activities, including in this case handicraft production and small-scale exchange, is not a new reality in the Andes, and likely not in many parts of rural Latin America. Pottery production and exchange has proven to be a useful livelihood activity in different social and economic contexts. What served as a viable strategy for single mothers during the hacienda regime is now being integrated into livelihood strategies for households with limited capacities in a neoliberal context. This raises the question of how the same activities might be adapted to different conditions, and highlights the importance of paying close attention to how these livelihood practices actually work, in the real places where they are used, as well as their change over time. The implementation of state or NGO-led projects with sustainable development or conservation goals, including payments for ecological services or climate change mitigation, perhaps represents a new context in which traditional livelihood activities will be adapted. Understanding the local logic of these activities is likely crucial to the outcome of such projects and schemes.

In this sense, the detailed analysis in the previous chapters of the patterns of pottery exchange in Olleros and the local rules governing pottery exchange customs has proven crucial. Throughout this discussion of trade, the multisited nature of livelihoods has been demonstrated, with special attention given to the unique spatiality (spatialities) of exchange networks. The spatiality of the pottery networks in Piura is distinct from the spatiality of textile networks, and likewise distinct from the spatiality of activities such as migration to work at coca plantations to the east. Yet a single Olleros household may carry out all three of these activities simultaneously, creating a larger scale spatiality for its entire diversified livelihood. This broader spatiality will be unique to each household, although will likely exhibit similarities with those of its neighbors. Regional differences may also be significant. Pottery exchange networks in Olleros are different than pottery exchange networks in other parts of Piura, and in other parts of the Andes. The same may be true for any of the other components of a diversified livelihood. Close attention to the "dynamic and diverse" nature of rural livelihoods is key (Perreault 2009: 450).

So why make pots? The story of Lucila Chuquihuanga, a potter interviewed during the 2004 fieldwork season explains it well. When asked how long she had been practicing this craft, she paused, considered her answer, and said: "It must be ten years, because when I had them (pointing to her children), I didn't make them, but then, from necessity…"[3] Lucila has no husband, pots were what she could do.

[3]"…harán pues, diez años de repente, porque yo teniéndolos no las hacía, pero después la necesidad…" (Lucila Chuquihuanga).

Pottery is never the only activity a woman or household will pursue, but it is one that can provide a remarkable degree of insurance against cash and land scarcity. Although its future as a viable livelihood activity may be uncertain, pottery production is an income source that has been available to Olleros women for generations, and is one that continues to be a significant component in their lives. Most simply, as the elderly potter Rosa Jiménez said, "With my pots I went about giving life."[4]

References

Brush SB (1980) Peru's invisible migrants: a case study of inter-Andean migration. In: Orlove BS, Custred G (eds) Land and power in Latin America. Holmes & Meier Publishers Inc, New York, pp 211–229

Flores Ochoa J, Nájar Vizcarra YD (1976) El Likira, intermediario ambulante en la cordillera de Canchas. Antropología Andina 1–2:125–135

Humphrey C (1985) Barter and economic disintegration. Man 20(1):48–72

Kay C (2008) Reflections on Latin American rural studies in the neoliberal globalization period: a new rurality? Development and Change 39(6):915–943

Larson B (1995) Andean communities, political cultures, and markets: The changing contours of a field. In: Larson B, Harris O, Tandeter E (eds) Ethnicity, markets, and migration in the Andes. Duke University Press, Durham, pp 5–53

Larson B (1998) Cochabamba, 1550–1900. Colonialism and agrarian transformation in Bolivia. Duke University Press, Durham

Mayer E (2002) The articulated peasant: household economies in the Andes. Westview Press, Boulder

McSweeney K (2004) The dugout canoe trade in Central America's Mosquitia: approaching rural livelihoods through systems of exchange. Ann Assoc Am Geogr 94(3):638–661

Orlove BS (1998) Down to earth: race and substance in the Andes. Bull Latin Am Res 17(2): 207–222

Orlove BS, Custred G (1980) The alternative model of agrarian society in the Andes: households, networks, and corporate groups. In: Orlove BS, Custred G (eds) Land and power in Latin America. Holmes & Meier Publishers Inc, New York, pp 31–54

Ortiz S (1980) The structure and regional articulation of dispersed rural settlements in Colombia. In: Orlove BS, Custred G (eds) Land and power in Latin America. Holmes & Meier Publishers Inc, New York, pp 129–152

Perreault T (2009) Environment and development. In: Castree N, Demeritt D, Liverman D, Rhoads B (eds) A companion to environmental geography. Wiley, Malden, pp 442–460

Ramón G, Bell MG (2013) Re-placing plainware: production and distribution of domestic pottery, and the narration of the pre-colonial past in the Peruvian Andes. J Anthropol Archaeol 32: 595–613

Reardon T, Vosti SA (1995) Links between rural poverty and the environment in developing countries: asset categories and investment poverty. World Dev 23:1495–1506

Reardon T, Berdegué J, Escobar G (2001) Rural nonfarm employment and incomes in Latin America: overview and policy implications. World Dev 29(3):395–409

[4]"Con mis ollas he ido dando vida señor" (Rosa Jiménez).

Salomon F (1981) Killing the Yumbo: a ritual drama of Northern Quito. In: Whitten NE Jr (ed) Cultural transformations and ethnicity in modern Ecuador. University of Illinois Press, Chicago, pp 162–208

Turner MD, Williams TO (2002) Livestock market dynamics and local vulnerabilities in the Sahel. World Dev 30(4):683–705

UNDP, UNOPS, SGP, FMAM, and IGCH (2004) "Las Ceramistas de Olleros" Promotional Poster

Valle L (1971) La ecología subjetiva como elemento esencial de la verticalidad. Revista del Museo Nacional, XXXVII, pp 167–196

Weismantel MJ (1988) Food, gender and poverty in the Ecuadorian Andes. University of Pennsylvania Press, Philadelphia

Appendix A
Profiles of the Potters and Traders Interviewed

M.G. Bell, *Pottery, Livelihoods, and Landscapes*, SpringerBriefs in Latin American Studies, DOI 10.1007/978-3-319-52331-6

#	Name	Potter (P), Non-Potter (NP), or Trader (T)	Age	Place of birth	Now lives	Children (y/n)	Spouse (y/n) Deceased (d)	Where learned?	When learned?	Who was teacher?	When started to Sell/Trade?	Taught daughters to pot?
1	Emilia Pintado Malacatos (& José Adán Troncos Bermeo)	P (& T, husband of P)	36 (& ?)	Culcapampa-Chinchinal (& Olleros-Aguayco)	Olleros-Cafetal	y	y	Olleros-Cafetal	16, after marriage	Mother-in-law (Eva Bermeo)	Before 19	No/not yet
2	Zenaida Huaman Chuquihuanga	P	30s/40s	Olleros-Sidro	Olleros-Pampa	y	y	Olleros-Pampa	18	Mother-in-law (Rosa Jiménez)	After 3 years practice	No/not yet
3	Micaela Jiménez Mondragón	P	53	Olleros-Aguayco	Olleros-Pampa	y	y	Olleros-Aguayco	12	Mother	?	Yes, but not practicing
4	Hortensia Criollo Jirón	P	50s	Olleros-?	Olleros-Cafetal	y	n	?	15	Mother	?	Yes, but still learning
5	Emma Mondragón Abad	P	50s	Olleros-?	Olleros-Cafetal	y	y	?	14	Mother, aunt	18	?
6	Orfilia Mondragón Mondragón	P	64/72?	Olleros-Aguayco	Olleros-Cafetal	y	y	Olleros-Aguayco	13	Mother	17, after marriage	Yes, 1 daughter practicing
7	Luzmila Parihuaman	P	75	Olleros-Callejon	Olleros-Pampa	y	y (d)	Callejon	18	Mother	?	No
8	Rosa Jiménez Parihuamán	P	75	Olleros-?	Olleros-Pampa	y	y (d)	?	10	Herself/ watching	?	Yes, 3 daughters-in-law & 1 step-daughter
9	Eva Bermeo	P	80	Olleros-Aguayco	Olleros-Cafetal	y	y	Olleros-Aguayco	?	?	?	Yes, 1 daughter-in-law
10	Oralia Mondragón Jirón	Ex-P and sometime T	60s/70s	Olleros-?	Olleros-Cafetal	y	y	?	12	Mother	20	No?
11	Irma Abad de Flores	Ex-P and sometime T	50s	Olleros-Aguayco	Olleros-Pampa	y	y	Olleros-Aguayco	12	Step-mother (Rosa Jiménez)	14	No
12	Altagracia Chuquihuanca	NP and T	50s	Culcapampa	Olleros-Pampa	y	y (d)					No
13	Honoria Garcia Parihuamán de Flores	NP and non-T	50s	Olleros-?/almost in Toronche	Olleros-Pampa	y	y					No
14	Beltrán Mondragón	T-son of P	30s/40s	Olleros-Cafetal	Olleros-Cafetal	?	y					
15	Mauro Mondragón	T-husband of P	68/77?	Olleros-Cafetal	Olleros-Cafetal	y	y					
16	Daniel Mondragón	T-son of P	60s/70s	Olleros-?	Olleros-Pampa	y	y					

Appendix B
Interview Questions and Topics

Interview Topics

Although written in the form of questions, these were used as prompts for the interviewer. All interviews were conducted in Spanish.

For potters and others who exchange/sell pots:

Where do you trade your pots?

For each location:

When do you go there?
How do you get there?

How long does it take to arrive?

How long is the whole trip?

Who do you trade with?

Did you know these people before? How?

What do you trade for?

What products do you want to get there?
What did you get last time you went there?
When are these products available?
What is the ratio used for exchange?

Tell me about the last time you took pots there.

When and where are regional markets?

For each market:

Do you take pots there?
Do you sell your pots or does another vendor?

M.G. Bell, *Pottery, Livelihoods, and Landscapes*, SpringerBriefs in Latin American Studies, DOI 10.1007/978-3-319-52331-6

Who is this vendor?

Do you sell your pots? Barter them?
How much/what do you get?
Who do you sell/trade your pots to/with?

Did you know these people before? How?

Tell me about the last time you took pots here.

When and where are regional festivals?

For each festival:

Do you take pots there?
Do you sell your pots or does another vendor?

Who is this vendor?

Do you sell your pots? Barter them?
How much/what do you get?
Who do you sell/trade your pots to/with?

Did you know these people before? How?

Tell me about the last time you took pots there.

Do customers come to your house for pots?

Do they bring products to trade?
Do they buy pots with cash?
Tell me about the last time a customer came to your house.

Where else can you get money for pots?

How long have you been making and exchanging pots?
How have things changed in your memory?

For consumers of pots:

Where do you usually get pots?
Do you prefer clay pots or metal pots?
Do you prefer pots from a specific place?
Do you prefer pots from a specific potter?
Do you prefer a specific type of pot?
Do you know any potters? How do you know them?

Do you trade for pots?

What crops/foods/products do you trade?
What crops do you grow? What animals do you raise?
When do the potters come to trade?
How much do you give per pot?

What are the exchange rates?

Tell me about the last time you traded for a pot.

Do you buy pots with money?

Where?
How much do you pay per pot?
Tell me about the last time you bought a pot.

What day and where are regional markets?
For each market:

Do you see potters/pots there?
Who sells the pots? Potters or other vendors?

Do you know the potters/vendors? How?

Do you buy/trade for pots there? How much does a pot cost?
Tell me about the last time you got a pot at this market.

When and where are regional festivals?
For each festival:

Do you see potters/pots there?
Who sells the pots? Potters or other vendors?
Do you know the potters/vendors? How?
Do you buy/trade for pots there? How much does a pot cost?
Tell me about the last time you got a pot at this festival.

How long have you been using and exchanging for/buying pots?
How have things changed in your memory?

Appendix C
Registry of Trade Routes, Destinations, and Customer Points of Origin

Complete registry of trading trip destinations recorded in interview data. Destination, type of "zone," approximate elevation range, distance (as described by traders), timing of trading trip, products exchanged, and number of times the location was mentioned in the interview data are all recorded. Note: If marked with * site was not possible to map.

Route 1: Southeast to the community of Matalacas

Destination	Zone	Elevation (m.a.s.l.)	Distance (one way)	When?	Products	# of references
Huasanche	Grain	3200–3600		August (best, but Aug-Nov)	Grains, peas, wheat, potatoes, maize (amarillo), broad beans	1
La Laguna*	Grain	n/a		August (best, but Aug-Nov)	Grains, peas, wheat, potatoes, maize (amarillo), broad beans	1
Maray (alto)	Grain	2800–3200	2 days	August (best, but July–Sept)	Peas, wheat	1
Matalacas (center)	Grain	n/a	2 days	August (best, but July–Sept)	Peas, wheat	1
Matalacas (sector unspecified)	Grain	n/a	2 days	August (best, but July–Sept)	Wheat, peas, barley, goats, poultry	2
Nangay	Grain	2400–2800	1½–2 days (3–4 days round trip)	August (best, but July–Sept), harvest	Peas, wheat, maize, hens, cheese	3

(continued)

M.G. Bell, *Pottery, Livelihoods, and Landscapes*, SpringerBriefs in Latin American Studies, DOI 10.1007/978-3-319-52331-6

(continued)

Destination	Zone	Elevation (m.a.s.l.)	Distance (one way)	When?	Products	# of references
Pedregal	Grain	3000+	1½–2 days (3–4 days round trip), 15 h (12am–5pm)	September, harvest	Grains, wheat, peas, barley, goats, poultry	4
Pircas*	Grain	n/a	2 days	August	Wheat, peas, beans	1
Las Rosas*	Grain	n/a	15 h (12am–5pm)	n/a	Grains	1
Salvia	Grain	2800–3200	2 days (3–4 days round trip)	July–Sept, harvest	Wheat, broad beans, peas, beans, grains, potatoes, maize (amarillo), poultry, barley, goats	4
San Antonio	Grain	2800–3200	1½–2 days (3–4 days round trip)	August, harvest	Peas, wheat, beans	2
San Carlos*	Grain	n/a		August (best, but Aug–Nov)	Grains, peas, wheat, potatoes, maize (amarillo), broad beans	1
San Jose	Grain	2400–2800	2 days, 15 h (12am–5pm)	August (best, but July–Sept)	Grains, peas, wheat, maize, hens, cheese, broad beans, barley, goats, poultry, potatoes, beans	5
San Juan	Grain	2400–2800	2 days	August (best, but Aug–Nov)	Grains, peas, wheat, potatoes, maize (amarillo), broad beans, beans	2
San Miguel	Grain	2800–3200	2 days, 15 h (12am–5pm)	August (best, but July–Sept)	Grains, wheat, peas, barley, goats, poultry	2
Santa Maria	Grain	2800–3200	3 or 4 days round trip	n/a	Wheat	1
San Pedro	Grain	2800–3200	3 or 4 days round trip		Grains, wheat	1
Yerbas Buenas	Grain	2400–2800	1½–2 days (3–4 days round trip)	Harvest	Peas, wheat	1

Route 2: Passing El Aypate: East to the communities beyond Mt. Aypate

Destination	Zone	Elevation (m.a.s.l.)	Distance (one way)	When?	Products	# of references
Ambulco	Grain	2400–2800	n/a	n/a	n/a	3
Carrizal and Sacalla	Grain	2000–2400	2 days, 3 with exchanges	n/a	Chickens, beans, maize, peas	4
Culcapampa	Grain	2000–2400	3 h, 1 day round trip	n/a	Beans, maize, peas	5
Granadillo	Grain	n/a	n/a	n/a	n/a	1
Lagunas de Canly	Grain	2400–2800	8 h	n/a	Barley, cheese, peas, wool	1
Simbaca	Grain	2400–2800	n/a	n/a	n/a	1
Singoya	Grain	2000–2400	1½ h	n/a	Grains	2
Talal	Grain	2400–2800	1 day	n/a	Beans, maize, olluco, peas, wool	4
Tapal	Grain	2000–2400	Pre-dawn to 9 am	n/a	Beans, chickens, goats, maize, olluco, peas	5
Tapal Alto	Grain	2800–3200	n/a	n/a	Grains	1
Vircales	Grain	2800–3200	9–11 h	n/a	Maize, olluco, peas, potatoes, wheat	3

Route 3: Lowlands to the south: Santa Rosa and Yanta

Destination	Zone	Elevation (m.a.s.l.)	Distance (one way)	When?	Products	# of references
Santa Rosa	Animals	1600–2000	½ day	n/a	Cheese, chickens, sweet potatoes	2
Yanta	Animals	1200–1600	n/a	n/a	Chickens, goats, pigs	1
Frejolito	Animals	n/a	n/a	n/a	Goats, menestras	1

Route 4: Passing Ayabaca: The city of Ayabaca and surrounding communities

Destination	Zone	Elevation (m.a.s.l.)	Distance (one way)	When?	Products	# of references
Ayabaca	Urban	3200–3600	4–5 h	n/a	Market products, money	4
Ambasal	Grain	1600–2000	n/a	n/a	n/a	1
Chocan	Grain	2000–2400	2 days	n/a	Lima beans, maize, peas, wheat	1
Los Huabos	Grain	2800–3200	n/a	n/a	n/a	1
La Laguna*	Grain	n/a	n/a	n/a	n/a	1
Pingola	Grain	2800–3200	3 h, or in car	n/a	Beans, broad beans, maize, peas, wheat	3

Route 5: Within and directly surrounding the community of Olleros

Destination	Zone	Elevation (m.a.s.l.)	Distance (one way)	When?	Products	# of references
Olleros-Congoli	Recado	n/a	2 h	n/a	Choclos (corn on the cob), guineo, manioc	2
Cujaca	Recado	1200–1600	<1 h	n/a	Manioc	1
Olleros-Toronche	Recado	n/a	2 h	n/a	Guineo, manioc	1

Others, unknown route (not mapped)

Destination	Zone	Elevation (m.a.s.l.)	Distance (one way)	When?	Products	n
Los Altos	n/a	n/a	n/a	n/a	n/a	1
Toldo	n/a	n/a	n/a	n/a	Cheese, peas, wheat, wool	1

Customer Points of Origin

Place of origin	# of References	Products
Ambasal	1	Cacao, chancaca
Ambulco	3	
Anchala	2	Alforjas, grains, menestras, wool
Aragoto	1	
Arraypite	1	Grains, menestras
Cuchayo*	3	Chickens, grains, guinea pigs
Cujaca	2	
Hualcuy	1	
Huanta	3	
Jilili	1	
Lagunas de Cauli	3	Grains, menestras
Matalacas (sector unspecified)	3	Peas, wheat
Montero	3	
Ollería	1	
Pacaipampa	1	Money
Pingola	1	Grains, menestras
Pintado	1	
Samanga/ Samanguilla	3	Grains, menestras
Sausal de Culucan	1	
Sicchez	2	Bananas, chancaca, manioc, money
Sichulqui	1	
Simbaca	1	
Sochabamba	2	Beans, grains, peas, money
Suyupampa	8	Beans, chickens, fish, grains, kerosene, maize, money, olluco, onions, peas, rice, salt, soap, wheat
Tacalpo	2	Grains, menestras
Talal	1	Beans, maize, peas
Tapal	1	Chickens, grains, guinea pigs

Appendix D
Registry of Products, Places, and Exchange Rates

Complete registry of products, places, and exchange rates recorded in interview data. Location specification (there/home) indicates whether the exchange is carried out at the trade destination or in Olleros.

Product	Place	Rate	Location (there/home?)
Bananas	Ambasal		Home
Bananas	Sicchez		Home
Barley	Lagunas		There
Barley (for chicha)	Matalacas		There
Beans	Culcapampa		
Beans	Matalacas	Potfuls	There
Beans	Matalacas		There
Beans	Sacalla	Half potfuls	There
Beans	Santa Rosa		There
Beans	Socchabamba		Home
Beans	Suyupampa		Home
Beans	Talal		There
Beans	Talal		Home
Beans	Talal	Potfuls in harvest, filled to a lower level not in harvest	There
Beans	Tapal		There
Beans	Tapal		There
Broad Beans	Chocan		There
Broad Beans	Matalacas	Potful	There
Broad Beans	Pingola		There

(continued)

M.G. Bell, *Pottery, Livelihoods, and Landscapes*, SpringerBriefs in Latin American Studies, DOI 10.1007/978-3-319-52331-6

(continued)

Product	Place	Rate	Location (there/home?)
Caballas (Fish)	Suyupampa		Home
Cacao	Ambasal	Calculate price of 1 libra of cacao and price of the pot	Home
Chancaca	Ambasal		Home
Chancaca	Sicchez		Home
Cheese	Lagunas		There
Cheese	Many places		Home
Cheese	Matalacas	1 cheese of about 1 kilo for 1–2 pots depending on size	There
Cheese	Santa Rosa		There
Cheese	Toldo	1 cheese = 2 pots or 1 tiesto or 1 cazuela	There
Coffee Beans	Olleros or Cujaca	potful	home
Goats	Frejolito	1 arroba of meat = 7 pots	
Goats	Matalacas		There
Goats	Matalacas	1 goat = 1 carga of pots or 1 arroba of chancaca	There
Goats	Santa Rosa	1 arroba of meat = 7 pots	there
Goats	Santa Rosa		There
Goats	Tapal	1 arroba of meat = 7 pots	There
Goats	Yanta		There
Grains	Matalacas	Potful	There
Grains	Matalacas		There
Grains	Suyupampa		Home
Grains	Suyupampa		Home
Grains	Tapal		There
Grains	Tapal	Potful	There
Guinea Pigs	Many places		Home
Hens	Matalacas		There
Hens	Santa Rosa	1 hen = 2 pots	There
Hens	Santa Rosa		There
Hens	Suyupampa		Home
Hens	Tapal	1 regular pot (1–2 lata capacity) =1 hen, 1 small pot (5 liters) =1 small chicken	There
Hens	Yanta		There

(continued)

(continued)

Product	Place	Rate	Location (there/home?)
Kerosene	Suyupampa		Home
Maize	Carrizal	Potful	There
Maize	Chocan		There
Maize	Culcapampa		
Maize	Culcapampa		
Maize	Many places		Home
Maize	Matalacas		There
Maize	Matalacas	Potful	There
Maize	Pingola	Potful	There
Maize	Pingola		There
Maize	Sacalla	Potful (?)	There
Maize	Suyupampa		Home
Maize	Talal		There
Maize	Talal	Potful	Home
Maize	Talal	Potfuls in harvest, filled to a lower level not in harvest	There
Maize	Tapal		There
Maize	Tapal		There
Maize	Tapal	Potful	There
Maize	Vircales		There
Manioc	Ambasal		Home
Manioc	Cujaca	1 regular pot = 1 arroba, 1 small pot = ½ arroba	
Manioc	Sicchez		Home
Menestras	Frejolito		There
Money	Ayabaca		There
Money	Socchabamba		Home
Money	Suyupampa		Home
Olluco	Talal		There
Olluco	Tapal	Potful	There
Olluco	Tapal	Heaped (*flotando*)	There
Olluco	Vircales		There
Onions	Suyupampa		Home
Peas	Carrizal	Half potfuls	There
Peas	Chocan		There
Peas	Culcapampa		There
Peas	Lagunas		There
Peas	Many places		Home
Peas	Matalacas		There
Peas	Matalacas	Potfuls	There

(continued)

(continued)

Product	Place	Rate	Location (there/home?)
Peas	Matalacas	Potfuls	There
Peas	Matalacas		Home
Peas	Matalacas		There
Peas	Matalacas		There
Peas	n/a	Lower down on the pot, at the base of the pot's "shoulder"	Unspecified
Peas	Pingola		There
Peas	Sacalla	Half potfuls	There
Peas	Socchabamba		Home
Peas	Suyupampa		Home
Peas	Suyupampa		Home
Peas	Talal		There
Peas	Talal		Home
Peas	Talal	Potfuls in harvest, filled to a lower level not in harvest	There
Peas	Tapal		There
Peas	Tapal		There
Peas	Tapal	To the pot's neck or shoulder	There
Peas	Toldo		There
Peas	Vircales		There
Pigs	Santa Rosa		There
Pigs	Yanta		There
Potatoes	Matalacas	Potful	There
Potatoes	Matalacas	Potful	There
Potatoes	Vircales		There
Poultry	Matalacas		There
Rice	Suyupampa		Home
Salt	Suyupampa		Home
Soap	Suyupampa		Home
Sweet Potato	Santa Rosa		There
Wheat	Chocan		There
Wheat	Many places		Home
Wheat	Matalacas		There
Wheat	Matalacas	Potful	There
Wheat	Matalacas	Potful	There
Wheat	Matalacas		Home
Wheat	Matalacas	Potful	There
Wheat	Matalacas		There
Wheat	Matalacas		There

(continued)

(continued)

Product	Place	Rate	Location (there/home?)
Wheat	Pingola	Potful	There
Wheat	Pingola		There
Wheat	Sacalla		There
Wheat	Suyupampa		Home
Wheat	Toldo		There
Wool	Many places		Home
Wool	Matalacas		There
Wool	Santa Rosa		There
Wool	Talal		There
Wool	Toldo	1 libra of wool = 1 large pot	There

Index

A
Agrarian Reform, 4, 11, 31, 38
Agricultural calendar, 13, 18, 46
Agricultural cycle, 18, 46
Andean Studies, 1, 2, 5
Animal, 2, 28, 33, 36, 43, 45, 46, 48, 51, 55
Argentina, 58
Ayabaca, 11, 28, 33, 36, 38, 41, 42, 45
Ayacucho, 62
Aypate, 32, 34, 55

B
Barter, 2–4, 8, 25, 27, 30–32, 37, 38, 44, 50, 51, 53, 57, 61–65
Beans, 27, 29, 30, 32–34, 36, 38, 45, 48, 49, 51–54
Bolivia, 11, 25, 26, 53, 58
Bulk purchase, 25, 27, 36, 38, 45

C
Cacao, 36, 45, 49, 55
Campesina, 11, 20, 27, 51. *See also* campesino
Campesino, 3–5, 8, 11, 20, 27, 32, 51
Cash, 3, 8, 32, 40, 42, 43, 57, 62–65, 67
Ceramic, 8, 51, 52, 62, 65
Chancaca, 17, 20, 28, 30–32, 36, 45, 51
Cheese, 45, 55, 56
Chicken, 18, 36, 38, 45, 54, 55
Choclo, 34, 35, 45
Chulucanas, x
Clay, 7–9, 13, 32, 52, 53, 65
Coca, 2, 18, 47, 49, 66
Coffee, 11, 13, 18, 45, 49, 51, 53
Community, 2, 11, 13, 18–21, 25–28, 30, 32–36, 38, 40, 43, 45, 46, 59, 61–63
Cooking pot, 9, 13, 52
Cuisine, 43, 50, 61
Customer, 20, 21, 25, 27, 36–38, 40, 45, 52, 54, 58
Cuzco, 25, 49, 55

D
Daughter, 16, 35, 65
Development Studies, 1, 5
Diversification, 1, 5–7, 61, 66
Donkey, 18, 27, 53
Dry season, 13, 27

E
Ecological complementarity, 1, 2, 7, 43, 46, 50, 51, 61, 62
Ecological zone, 2, 5, 7, 11, 46, 47, 50, 57, 61. *See also* zone
Ecuador, 63
Ethnoarchaeology, 9, 63
Exchange, 1–5, 7, 8, 10, 16, 19, 21, 25–28, 30, 32, 35, 36, 40, 42, 43, 45–47, 50–59, 61–66
Exchange rate, 4, 27, 43, 52–55, 57, 58, 65, 66
Ex-hacienda, 11, 30, 33. *See also* Hacienda

F
Fair, 25–27, 31, 39
Farm, 1, 3, 5, 6, 8, 43, 65
Fieldwork, 10, 18–20, 34, 36, 66
Food, 5, 6, 8, 13, 18, 34–36, 40, 43, 45, 47, 50–52, 55, 61, 62, 64, 65

G
Girl, 16, 58, 59
Goat, 33, 45, 48, 55
Grain, 18, 27–33, 36, 43, 45–48, 50, 52–58, 65
Guinea pig, 36, 38

M.G. Bell, *Pottery, Livelihoods, and Landscapes*, SpringerBriefs in Latin American Studies, DOI 10.1007/978-3-319-52331-6

H
Hacienda, 11, 30, 31, 38, 64–66
Handicraft, 7, 51, 62, 66
Harvest, 2, 18, 27, 28, 30, 34, 46, 51, 54, 56, 58
Hen, 55, 82. *See also* chicken
Household, 4–9, 16, 18, 28, 43, 50, 51, 54, 58, 59, 61, 62, 64–66
Huarmaca, 9. *See also* Santa Ana de Huarmaca
Husband, 19, 27, 58, 64, 66

I
Indian, 64
Interview, 2, 18–20, 28, 29, 33, 34, 36, 37, 43–45
Itinerant potter, 40, 41

K
Kerosene, 32, 36, 45

L
Landscape, 2, 4, 5, 40, 46, 50, 59, 63, 66
Life cycle, 20, 43, 58
Livelihood, 1, 2, 3, 5–8, 18, 20, 25, 32, 43, 47, 49, 59, 61, 62, 64–67
Livestock, 5, 7, 18, 31, 33, 48, 64

M
Maize, 2, 8, 9, 11, 16, 18, 27, 29, 32, 33, 36, 45–47, 49, 50, 52–54, 56
Manioc, 16, 34, 36, 44–46, 49, 50, 55
Map, 21, 25
Market, 3–8, 25, 27, 31–33, 36, 38, 41, 42, 45, 49, 51, 58, 61, 64, 65
Matalacas, 28–32, 45, 51, 53, 54
Measurement, 3, 4, 43, 51–53, 56–59, 66
Meat, 50, 55
Menestra, 36, 43, 45, 50
Metal, 13, 53, 65
Money, 3, 8, 13, 27, 28, 30, 36, 38, 43, 45, 56, 57, 61, 62. *See also* cash

N
Nepal, 57
Network, 19, 21, 25, 51, 54, 55, 59, 62–66
New Rurality, 1, 5, 6
Nonfarm activity, 6, 8, 64

O
Olleros, 1, 2, 4, 8, 9–21, 25, 26, 28–42, 43–59, 61–67, 70, 82. *See also* San Bartolomé de los Olleros
Olluco, 45, 50, 53

P
Peanut, 16, 18
Peas, 11, 27, 29, 32, 33, 36, 38, 44, 45, 50, 51, 53, 54
Peonage, 18, 20, 40
Piura, 2, 8, 9, 27, 28, 38, 43, 44, 47, 48, 50–52, 62, 66
Plantains, 16, 34, 45
Plastic, 53, 65
Political ecology, xi
Potato, 2, 4, 29, 44, 45, 47, 49, 50, 53
Potful, 28, 52, 54, 55, 57, 65
Pots and Potters, 1, 2, 6–11, 13, 16, 18–21, 25–38, 40, 42, 43, 45, 46, 48–59, 61, 62, 64–67
Poultry, 44
Poverty, 5, 6, 8, 20, 61, 64

R
Recado, 34, 43–45, 48, 50

S
San Bartolomé de los Olleros, 1, 2, 4, 8–11, 13, 16, 18, 19, 21, 25, 26, 28–36, 38–40, 42–46, 48, 51–54, 56, 57, 59, 61–67
San Juan Bautista de Sondorillo, 9, 10
Santa Ana de Huarmaca, 9
Santo Domingo, 9
Seasonality, 3, 53
Sierra, 9, 28, 38, 45, 47, 48, 50, 54, 63
Single mother, 58, 65, 66
Sondorillo, 9, 10. *See also* San Juan Bautista de Sondorillo
Spatiality, 25, 43, 59, 61, 66
Sugarcane, 16, 18, 32, 36, 47, 64
Sweet Potato, 44, 50, 51

T
Temple, 30, 45
Textiles, 18, 36, 62
Toasting pot, 9, 13, 53
Tools, 13, 21
Trade, 2–4, 7, 8, 19, 21, 25–28, 31–33, 35, 36, 38–41, 43–46, 49–55, 57–59, 61–64, 66
Trade route, 25, 45, 46, 54, 55
Traders, 3, 16, 18, 19, 21, 27–34, 37, 38, 43–46, 48–51, 53, 57, 58, 62, 63, 65
Trading trip, 19, 20, 25, 27–29, 32, 35, 36, 38, 40, 54, 58, 65
Trapiche, 18, 64
Traveling trader, 45
Tuber, 4, 45, 47–49, 52

U

United Nations Development Program (UNDP), 51, 65
Urban market, 33, 41

V

Vertical archipelago, 4, 7, 46, 50
Verticality, 2, 7, 46, 50. *See also* vertical archipelago

W

Wheat, 9, 11, 27–29, 31–33, 36, 44–46, 49, 51–54
Women, 9, 13, 16, 18, 25, 27, 36, 41, 64, 65, 67
Wool, 29, 36, 45, 55, 56
Workshop, 9

Z

Zone, 2, 4, 5, 7, 11, 18, 28, 30, 32, 33, 45–50, 54, 57, 61

MIX
Papier aus verantwortungsvollen Quellen
Paper from responsible sources
FSC® C105338

If you have any concerns about our products,
you can contact us on
ProductSafety@springernature.com

In case Publisher is established outside the EU,
the EU authorized representative is:
Springer Nature Customer Service Center GmbH
Europaplatz 3, 69115 Heidelberg, Germany

Printed by Libri Plureos GmbH
in Hamburg, Germany